AI for Real Companies

A Practical Guide to Smarter Systems and Stronger Profits

AI for Real Companies

A Practical Guide to Smarter Systems and Stronger Profits

Deb Weidenhamer

Presented by AI Business Advisory Institute

An Imprint of AI Business Advisory Institute Publishers

New York, NY

www.aibusinessadvisoryinstitute.net

Library of Congress Cataloging-in-Publication Data

Names: Weidenhamer, Deborah, author.
Title: AI for Real Companies: a practical guide to smarter systems and stronger profits / Deborah Weidenhamer
Description: New York, NY. AI Business Advisory Institute, 2026.
Identifiers: LCCN 2026903327 | ISBN 979-8-9947971-0-5 (paperback)
Subjects: LCSH: Artificial intelligence—Business applications. | Business enterprises —Technological innovations | Digital transformation | Organizational effectiveness.

CONTENTS

AI for Real Companies

Introduction

Smart business owners are not losing sleep over whether AI works. They are lying awake because they know it works and are worried a competitor will steamroll their company before they figure out AI.

You may be one of those entrepreneurs, someone who has built a successful company through grit, **customer relationships**, and plain old hard work. Now you suddenly feel like you are falling behind in a race you never chose to enter.

You find yourself caught between two equally terrifying possibilities: spending money you do not have on **technology** you do not understand, or watching your competition pull ahead while you are still trying to figure out **what "machine learning" actually means.**

We get it. You did not start your business to **become a tech expert.** You started a company because you are good at what you do, whether that is managing a law firm, running a restaurant, or selling custom-made furniture. Now, everyone is saying that if you are not using AI by tomorrow, you will be obsolete by next month and out of business by the end of the year.

Here is what those “know-it-alls” are leaving out: **artificial intelligence is not some mystical force** that only works for big corporations with unlimited budgets. It is a tool. And like any tool, it works best when you understand what it is actually supposed to do for your business.

The AI Business Advisory Institute

Enter the **AI Business Advisory Institute.** We are not a think tank full of academics theorizing about the future of work. **We are AI Engineers, AI Policy Managers, and AI Architects** who collaborate as experienced business consultants. We assist companies in implementing effective strategies and adapting their skills to address developments in **artificial intelligence** and its impact across industries. Our team knows the difference between impressive-sounding technology and technology that pays for itself within six months.

We translate complex systems into precise results because we have learned something important. The best **AI implementation** happens when you keep the human touch that makes your business thrive. You do not need to choose between embracing technology and maintaining the **personal connections** that contributed to your success. Learn more at AIinstituteforBusiness.net.

This book will show you exactly how to **harness AI** without losing what makes you different. You will learn to **work smarter, serve customers better,** and sleep more at night while your competition is still trying to find their footing.

Why This Voice Matters

Deb Weidenhamer, AI Engineer and lead author of this book, has spent her career guiding businesses to stronger through **resilience, technology,** and clear **leadership.** A former New York Times business contributor and founder of an INC 500 company, she has built and led teams that help other businesses improve their **systems** and **profitability.**

Like many **entrepreneurs** who take bold risks for their companies, she faced her most challenging test when a business decision led to a short stay in a federal prison camp. The same kind Martha Stewart once visited. That experience gave her an unfiltered understanding of **accountability,** resilience, and the real cost of leadership. Deb rebuilt her career, using those lessons to help other business owners strengthen their companies, protect what they have created, and approach innovation with both courage and caution.

Why You Will Relate

You established and grew your business through long hours, hustle, and persistence, handshakes, **personal relationships,** and a reputation that took years to develop. Your customers know your name, trust your judgment, and value the **individual attention** you provide.

Now you have an opportunity to **enhance everything you have created** with tools that highlight your strengths. Other business owners recognize **AI's business potential**. Your biggest competitors are using artificial intelligence to steal customers you do not even know you are losing. Every day you wait to bring AI into your business, a competitor is moving faster, learning more, and widening the gap. The longer you delay, the more complicated and expensive it becomes to catch up.

While you are reading this book, **tech-savvy companies** are methodically dismantling businesses just like yours using **artificial intelligence tools** that many traditional business owners have never even heard of.

They are not replacing people relationships; they are using AI to make their human contacts more valuable than yours. They deliver the **personal attention you provide**, but faster and more consistently, with insights from analyzing thousands of **customer interactions** rather than relying on memory and intuition alone.

The Truth About Competitive Disruption

Disruption happens gradually, then suddenly. One month, you are the established leader in your market. The next month, you wonder why customers are choosing a competitor that delivers better service at lower prices.

After helping many businesses implement **process automation** and **business performance improvements** for more than fifteen years, moving into artificial intelligence was natural for us. We have watched business owners who wait "until next quarter" to evaluate AI watching their **market share** disappear to competitors who started their AI journey six months earlier.

The window for your **strategic advantage is narrowing** by the day.

Personal Connections - Your Secret Weapon

What big business wants to hide from you is that your **personal customer connections** are not a weakness. Your relationship network is your **secret weapon** waiting to be unleashed.

The same AI that tracks Amazon's millions of customers will help your neighborhood mechanic remember that your 1988 Suzuki Samurai burns oil faster than usual, automatically bumping up routine maintenance reminders by three weeks.

A Phoenix bakery owner can implement **inventory management AI** for under $200 per month and see results within weeks. A Denver plumber can use **predictive maintenance AI** to prevent shop equipment failures while still personally fixing water heaters with the same care that built his reputation.

You do not need to hire programmers or spend months learning software. Our research from working with businesses shows you do not need to become a tech company to succeed. **AI helps you become a more powerful version of yourself.**

A bakery owner does not need to **understand algorithms** any more than he needs to understand combustion engines to drive his delivery truck.

Using This Book As a Reference and Guide

This is not a technology book; **it is a business book about technology.** It is packed full of **success stories** about real companies that have successfully implemented AI in their business and true tales of AI failures, because without proper planning, those happen too.

Broken into three sections, the book opens with **Part 1: TRANSFORM**—a primer on artificial intelligence and its vocabulary that takes AI from a place of fear and confusion to a strategic business opportunity. **Part 2: TRACK** is a proven methodology that ensures successful AI implementation within your company. Wrapping up with **Part 3: INNOVATE** offering **industry-specific AI application recommendations and actionable guidance** for different types of businesses, including how to implement AI successfully, based on the successes, lessons, and timelines of similar companies.

AI Tools

Throughout the book, you will find specific references to AI Tools. AI tools are programs designed for specific industries and applications. ***AI Tools*** are indicated in italics and bold. Current pricing is listed. At the end of this book, there is an appendix of all the AI tools listed in this book.

New AI tools come to market every day, so always do your research. New additions to the market are not just something we say about **AI tool releases**. This is a new world where new tools enter the market every few weeks. **AI learns 4% new information every week**, so imagine how that can impact an industry and a business.

A Note on Pricing and Tool Selection

To help you budget and plan, throughout this book, you will find pricing information for various AI tools. These prices were current at the time of writing, and are provided to help you budget and plan for your AI implementation. However, **software pricing changes frequently,** often several times per year.

The ranges shown reflect typical costs for **small-to-medium businesses** using standard configurations. Your actual costs may vary based on the number of users, locations, features, modules, and other discounts available from the manufacturer or through business associations specific to your industry.

The pricing in this book serves as a **planning guide**, not a final quote. Focus on the **implementation strategies, use cases,** and **ROI calculations**. The tactical approaches remain valuable long after prices change.

Similarly, **AI capabilities** evolve rapidly. A tool mentioned here may have added features, changed names, or been acquired since publication. Use the company names and strategies as your starting point, then verify current offerings.

The most expensive tool is not always the best fit, and the cheapest option may cost you more in lost productivity. Choose based on business needs, not just price tags.

Success Stories

Throughout this book, you will find Success Stories—real examples of businesses that applied the **framework to achieve measurable results** with AI. These stories go beyond theory. They show what business owners did, which AI tools they used, the solution cost (at the time of writing), and the outcomes they achieved. Each story illustrates how **technology, team development,** and **customer engagement** work together to create lasting **competitive advantages**.

A **Success Story** is not about a billion-dollar tech firm or an abstract innovation. These tales are of business leaders, just like you, who combined **practical tools** with thoughtful leadership to turn AI into a genuine **growth engine.** Their results prove that AI mastery is not about size or budget. It is about mindset, method, and timing.

In some cases (especially failures), we use only broad business descriptions and the fictitious names of business owners. There will be times when an asterisk (*) follows a name to indicate it is a substitution for the business or business owner/manager's true name.

When using company **case studies**, you will see the company name in *italics*. By visiting the website of this book, you will find a full appendix of all the company case studies.

Additional **tools, assessments, templates**, and other items you may find useful during your AI implementation can be found at our website: **AIforRealCompanies.com.**

Part I: Transform

Chapter 1

AI Demystified

It is 3:17 a.m. again. You are lying awake, laptop in hand, watching your competitor's **chat widget** respond in real time while your website does nothing. Your competition added "AI-Powered" to their tagline six months ago. Since then, three of your best customers have mentioned how impressed they were with their new service.

The math keeps you awake. If they pinch one customer per week, that adds up to more than 50 customers in a year. At a $5,000 average customer value, that is a quarter million dollars leaving your door. While you struggle to figure out what **"machine learning"** means.

Your dog sighs in annoyance, clearly bothered that you are disturbing his sleep. You close the laptop, but the questions remain. How long before all your customers discover what you already know—that someone else is doing what you do, only faster and cheaper?

The AI Opportunity

We are standing at the edge of one of the most exciting times in modern history, perhaps the biggest shift in the last 150 years. **Artificial intelligence** is not just another **technology trend**; it is the new language of **competition**. And right now, regular businesses finally have a chance to go toe to toe with the giants and even outsmart them.

The real breakthrough is not only about **efficiency**, but about amplification. Artificial intelligence allows you to do what you have always done best: know your customers personally. But now, you can pair that human insight with the kind of **business intelligence** that once required entire departments and million-dollar software systems. You can **predict customer needs, optimize inventory**, and **coordinate operations** with the precision of a Fortune 500 company. Large competitors can match your scale, but they will never match your connection.

The Competitive Reality You Cannot Ignore

A US Chamber of Commerce survey of more than 1,000 businesses shows the market shift toward artificial intelligence is **real** and **accelerating**.

AI Adoption Facts:

- 40% of businesses used generative AI, nearly doubling from 23% in the previous year.
- 86% of business AI users say **AI improved operational efficiency**.
- 47% fall into "high" or "very high" **tech adoption** categories.

The Most Common AI Applications

AI applications exist for specific industry use and for common use across all companies. These include **artificial intelligence** for:

- Marketing and promotions: (***Hubspot; Surfer SEO***) 56%
- Customer communication: (***tidio; Drift***) 51%

- Customer insights and engagement trends: ***(Crescendo; Salesforce Einstein)*** 51%
- Employee recruitment: **(*Indeed; Workday*)** 32%

Your Competitors Are Moving Fast

Forward-thinking competitors use sophisticated **automation tools**. They implement **AI systems that predict customer needs** before customers know they have them. They **personalize communications**, **optimize operations** in real time, and deliver memorable experiences that highlight both **efficiency** and **personal attention**.

While some businesses manually track customer preferences, these leaders use **machine learning algorithms**. Systems that remember every interaction, purchase pattern, and communication preference across thousands of customers simultaneously.

What is Artificial Intelligence?

Before you can use AI to your advantage, you must speak its language. This chapter breaks down the words, phrases, and acronyms that everyone throws around, so you can join the conversation, spot the hype, and use the right terms to create real results in your business.

AI is not mystical technology destined to rule the world. It is sophisticated **pattern recognition**—like Sesame Street's "some of these things are just like the others," except AI processes thousands of patterns instantly and improves each time it sees more information.

Simply put, AI **finds signals in data** that people would miss or take too long to recognize. No magic, no mystery, just powerful tools you control and direct.

Essential AI Categories

AI Assistants

AI Assistants and AI Tools referenced are shown in ***bold italics***.

ChatGPT, Microsoft Co-Pilot, Google Gemini, Claude, and ***Perplexity*** create new content and produce human-like answers based on vast training data. These assistants can author emails, create marketing copy, answer customer questions in a natural and conversational manner, perform calculations, organize data, create charts, and design visuals.

AI Applications

An **AI application** is software built around **artificial intelligence models** that are specific to an industry or a need. Each AI application has a specific niche or focus. Some analyze numbers to **forecast sales** or **inventory needs**. Others interpret language, images, or patterns to speed up creative work or **automate routine tasks**.

AI applications turn intelligent technology concepts into real-world solutions. These programs embed machine learning and natural language capabilities directly into everyday tools — from voice assistants like Amazon's ***Alexa*** and marketing platforms such as ***Einstein***, the Salesforce AI for customer CRM, to customer-service chatbots like ***Chatfuel***, and translation tools comparable to ***Google Translate***.

These tools can **predict outcomes, generate content**, or assist with decisions that once required human judgment. What defines these applications is not just automation, but the ability to **learn and adapt from data** over time.

Throughout this book, you will see dozens of examples of **AI applications** that have been engineered for specific small and medium-sized business categories. Also known as **AI-powered apps**, these tools help businesses **make smarter decisions, serve customers more efficiently,** and **compete** at a level that previously required entire departments of analysts.

AI Capabilities

Artificial intelligence has many forms, but every tool falls into one of a few functional categories, or AI capabilities. These describe what technology actually does.

- **Predictive AI** looks at past data to forecast future outcomes.
- **Generative AI** creates something new: words, images, code, or designs.
- **Visual AI** interprets what it sees: turning photos, video, or live feeds into useful insights.

Each capability serves a different purpose, but together they form the practical toolbox that businesses can use to operate **smarter, faster, and with more precision** than ever before.

Remember lying awake, wondering if you could afford that new hire? AI does not need health insurance**. AI does not take vacations.** And it remembers every customer interaction perfectly. One Phoenix bakery owner told us, "I spent three months agonizing over hiring a $15/hour assistant. Then I discovered AI could handle 80% of those tasks for $200 a month. I hired an assistant anyway to manage the AI and focus on customers. Best of both worlds."

Success Story
LL Flooring Solutions

LL Flooring in Richmond, VA took the AI visualization approach and flipped it for in-store use. Every sales advisor carries an iPad equipped with ***Roomvo AI***. Customers are encouraged to bring a photo of their room to the showroom, and the sales team uses the AI to explore different flooring options together.

The AI analyzes customer photos, automatically detecting floor surfaces, walls, lighting conditions, and room type. It then renders a realistic preview of how different flooring options would look in that specific space. **Machine learning** improves over time, getting better at handling tricky angles, unusual lighting, and partially visible floors.

"The store associates say it is a great closing tool," says Charles Tyson, *LL Flooring's* president and CEO. "Shoppers just love it since it takes the guesswork out of selecting the perfect floor for their space."

What makes this interesting is how it changes the sales conversation. Instead of talking abstractly about products, the salesperson and customer are looking at the same visualization together. They can swap options, compare colors, and make decisions collaboratively. **The AI becomes a tool that enhances the human relationship rather than replacing it.**

The technology also reduces returns. When someone can see exactly how a floor will look in their actual living room, they are less likely to be surprised when it arrives. That saves money on restocking, reduces waste, and keeps customers happy.

Verified Results:

- 3.7 times more **sales conversions** compared to traditional approaches.
- **Stockouts** decreased 78% while overall inventory levels dropped 23%.
- **Emergency rush orders** were reduced by 89%.
- **Inventory turns** improved from 4.2 turns to 6.8 times annually.
- **Sales increased 43%** in the first year due to better availability.
- By reducing rush orders, **profit margins improved 31%.**

AI Capabilities: **Predictive AI**

Predictive analytics use artificial intelligence to look backward so you can look forward. By analyzing patterns in past data such as **sales trends, customer activity,** and **inventory cycles,** AI identifies what is likely to happen next.

It transforms raw information into **foresight**, allowing you to **anticipate demand, adjust pricing,** or **spot potential problems** before they surface. In business terms, it is like having an experienced analyst on staff who never sleeps and always has the numbers straight. Tools such as ***QuickBooks AI*** and **customer behavior prediction platforms** use predictive analytics every day **to forecast cash flow,** recommend timing for inventory orders, and reveal which clients are most likely to buy again.

AI Capabilities: **Generative AI**

Creates new content such as **text, images, or music**. The content is based on patterns learned from millions of examples, like having a talented assistant who studied thousands of successful business letters and can now write professional correspondence that sounds authentically like your company.

AI Capabilities: **Computer Vision**

Technology that lets computers 'see' and understand images and videos like humans do. For business, this means software that automatically identifies **damaged products, counts inventory,** or recognizes when equipment needs maintenance just by looking.

AI Capabilities: **No-Code AI Platforms**

No-code AI platforms allow non-developers to build **AI-powered** tools without knowing how code is built or understanding coding languages. Examples: ***Zapier AI, Bubble,*** and ***Retool***.

AI Capabilities: **Natural Language Processing (NLP)**

Enables computers to **interpret and respond to natural human language**, eliminating the need for specialized commands or programming code. It enables users to ask business software questions in plain English and receive useful answers in return.

AI Capabilities: **Recommendation Platforms**

These AI models suggest content or products based on customer behavior. Examples: ***Netflix, Amazon,*** and ***Spotify***.

Applied AI Tools

Applied AI tools are where artificial intelligence gets down to business. They take the theories, models, and data science behind AI and turn them into something you can actually use. These are the programs, platforms, and connectors that make **AI practical** – writing emails, analyzing sales trends, answering customer questions, or syncing data between systems. Some are easy plug-ins that enhance tools you already use, like

QuickBooks AI or ***Canva***. Others are full platforms built entirely around AI, such as ***ChatGPT*** or ***Perplexity***. Whether the tools automate a simple task or manage complex workflows, applied AI tools are the hands and feet of artificial intelligence — the part that turns potential into performance.

Applied AI Tools: **Search + AI Combination**

Blends traditional search engines with **AI-generated answers**. Instead of just showing links, these systems interpret your intent and produce summarized, conversational responses. Examples include ***Bing AI*** and ***Google's Search Generative Experience (SGE).***

Applied AI Tools: **Chatbots**

Software that has **basic conversations with customers** through text or voice. Think of it as a patient employee who never tires of answering questions about store hours, return policies, or whether you are accepting new clients.

Applied AI Tools: **Industry-Specific AI**

Custom AIs designed for specific business types. Examples include ***Harvey*** (legal), ***Med-PaLM*** (healthcare), and ***Restb.ai*** (real estate).

Applied AI Tools: **Application Programming Interface (API)**

API's are how software applies artificial intelligence. These are behind-the-scenes connectors that enable your business software to integrate with **AI capabilities**. Think of APIs as electrical outlets for software. Just like plugging a laptop into the wall connects with electricity without understanding power grids, APIs let business systems connect with AI capabilities that other companies have built. APIs make it possible for tools like ***QuickBooks, Shopify,*** or ***HubSpot*** to use **AI features** built elsewhere.

Applied AI Tools: **Prompts**

Prompts are how people apply AI. Prompts are the **instructions or questions** you give an **AI system to** provide the results you want. Like directing an employee, the clarity and specificity of your prompt determine the quality of the work that comes back. Writing strong prompts is one of the most valuable business skills in the **AI era**. You can execute prompts in AI Assistants like ***ChatGPT, Perplexity***, and ***Claude***.

Core AI Terms

Algorithm: The information that computers follow to get things done. Like your grandmother's cookie recipe with specific steps, algorithms tell computers exactly how to complete tasks like "find the best route to three addresses" or "identify customers who have not been contacted in six months."

Machine Learning: AI that improves predicting things by analyzing patterns in your data. Instead of programming specific rules, the system figures out what works by studying examples and outcomes — much like an employee who gets better with experience.

Neural Network: AI designed to work somewhat like the human brain, with interconnected "neurons" processing information. These systems need a lot of data to learn patterns. Once trained, the networks can recognize speech, identify images, or predict customer behavior. Neural networks are the foundation for most modern AI, including today's **large language models**.

Large Language Models (LLM): AI is trained with an entire library of information with massive amounts of text, including books, articles, and websites. An LLM can understand and generate human-style writing. Think of accessing someone who has read almost everything written and can apply that knowledge to specific requests.

GPT (Generative Pre-Trained Transformer): A specific type of large language model that creates content by predicting what comes next in text. Trained in billions of words, GPT learns language patterns and uses that knowledge to generate clear, conversational writing. It is the engine behind tools like ***ChatGPT*** that can draft emails, summarize documents, or write code.

Natural Language Processing (NLP): Enables computers to interpret and respond to natural human language, eliminating the need for specialized commands or programming code. It enables users to ask business software questions in plain English and receive useful answers.

Model Drift: When your **AI system's performance** degrades over time, even though nothing has changed in your setup. You trained it on past data, but the world has changed and moved on. Customer behavior shifts, markets evolve, and your AI may still be making decisions based on yesterday's reality.

Hallucination: When AI systems give completely wrong information while sounding confident. It is like having an employee who makes up facts but presents them so convincingly that you believe them. Always double-check **AI outputs**, especially for important business decisions.

Predictive Analysis: The Crystal Ball For Your Business

Your **AI assistant** can predict what customers will buy next month, when equipment will break down, and which marketing campaigns will work—all before any of it happens.

Predictive analysis is not magic, though it might feel like it. It is AI looking at patterns in business data and making educated guesses about what is next. Think of it as having a super-talented marketing director who never forgets anything and spots trends you would never notice.

How Predictive Analysis Works

Your business generates data daily—when customers buy, what they purchase, seasonal patterns, weather impacts, local events, and other data that helps determine customer behavior. Most information just sits unused. **Predictive AI** takes historical data and looks for repeating patterns.

Simple example: every February, sales drop 20%. Whenever it rains three days straight, service calls increase 40%. Every time the local college has big games, weekend revenue jumps 15%. You might notice some patterns yourself, but AI spots hundreds of anomalies simultaneously and calculates exactly how they interact.

The Historical Pattern: Why AI Is Just Another Steam Engine

When **steam engines** were invented, the same basic technology powered textile mills, mining operations, trains, ocean liners, manufacturing plants, printing presses, flour mills, breweries, sugar refineries, sawmills, and early electrical generators.

What made steam engines transformative was exactly what makes AI powerful today—it was **general-purpose technology** enhancing worker capabilities across completely different industries. Textile weavers did not need a thermodynamics degree to benefit from steam-powered looms, just like today's restaurant owners do not need an education in **machine learning** to benefit from **AI inventory management**.

Steam engines succeeded by **enhancing existing expertise** rather than replacing it. Expert textile weavers still knew which patterns customers wanted and how to maintain quality standards. Steam power increased the value of weavers' skills by eliminating the physical limits of manual work.

The **businesses that thrived** during the steam revolution were not those abandoning traditional methods entirely. They utilized steam power to enhance what they were already skilled at, doing it better. The local blacksmith who added steam hammers did not stop being a craftsman—he became a craftsman, taking on bigger projects and serving more customers without sacrificing personal attention.

Consider small-town newspapers during the steam era. Before the advent of steam-powered printing presses, local papers typically printed only two hundred copies, a process that could take days to complete. Steam power enabled the same editor to print 2,000 copies in hours. But successful papers were not just printing more paper; they used expanded capacity to provide more thorough local coverage and better service to advertisers and subscribers who valued community connection.

The steam revolution demonstrated that **organizations that flourished were not those that rebranded themselves** as steam companies, but rather those that leveraged steam **technology to enhance their existing operations.** Similarly, companies adopting artificial intelligence are not transforming into **AI businesses**; instead, they are utilizing AI to strengthen and advance their core capabilities.

Chapter 2

Personal Relationships Your Secret Weapon

The greatest **competitive advantage** is not what you think it is. After helping many businesses implement **AI**, we have discovered something that surprises most business owners: while **tech-native companies** excel at **automation** and **efficiency**, they struggle with something that cannot be programmed—**authentic human connection**.

Your decades of **relationship-building** have not created a liability in the digital age. They have laid the groundwork for an **AI-powered market dominance** that tech companies cannot replicate. No matter how much venture capital they may raise or how many engineers they hire.

The businesses best positioned to thrive are not those that abandon **personal relationships** for tech efficiency, but those that use **artificial intelligence** to **amplify the very personal connections** that created their success.

Your traditional approach is not outdated—it is the **secret weapon** that transforms **AI** from a threat into an unstoppable **competitive advantage**. But only if you implement it correctly.

The False Choice

The **tech industry** promotes a false choice that we find personally frustrating: they claim that modernization requires abandoning everything that makes your business special. They want you to believe that AI adoption means replacing personal connections with cold, impersonal automation, trading personal service for efficiency, and **sacrificing business identity** for **technological capability**.

This is pure nonsense, and it has kept millions of established businesses hesitant to act while early AI adopters continue to grow.

The care and concern you have for employees and customers are not obstacles to AI adoption. They are exactly why you will ultimately lead the market once you implement AI thoughtfully.

While **tech-native businesses** excel at efficiency, they **struggle with human elements** that most companies have mastered. Tech businesses can automate transactions but cannot automate trust. They can personalize communications but cannot replicate authentic relationships established over decades. They can optimize processes but cannot create employee connections that turn customers into lifetime advocates.

Learn From Others' Mistakes: When AI Fails

Jewel * (starred names are used to blur the true identity of a person or company) built her accounting firm on trust. Twenty years of tax returns. Family secrets shared over coffee. When her largest client asked if she was "keeping up with all this AI stuff," her stomach dropped.

"I looked him in the eye and lied," she told us. "Said we were all over it. Then I went home and cried. **How could I tell clients who trust me with their financial lives that I don't even understand what ChatGPT is?"**

That conversation **happens every day** in businesses across America. Good people who built something real, feeling like frauds because they have not mastered technology that did not exist three years ago. Before we outline what works well with AI, let us show you what does not. **Failure is the best teacher**, especially when the failure is not yours.

Roland's Landscaping

*Roland's Landscaping** in Dallas spent $12,000 on an **AI scheduling and route optimization tool** that promised "revolutionary efficiency improvements". The company implemented the new technology without training employees or communicating with their customers.

The result: Customers complained about losing personal contact with their regular staff. Employees resisted using the new system because nobody explained how it helped them, and the AI tools could not handle the custom requests that made *Roland's* different from its competitors.

After eight months, the company abandoned the AI tools and returned to manual scheduling, having wasted money and damaged both employee morale and customer relationships.

Thompson Electronics – Austin

*Thompson Electronics** in Austin spent $400 a month on a chatbot for their website, provided no training to their staff, and informed customers of nothing. Within six weeks, 23% of their regular customers stopped calling. The AI did not fail. The rollout did.

Employees resisted, customers felt abandoned, and the company wasted $2,400 on technology while damaging relationships built over years that are difficult to price.

The difference between failure and success is not the quality of AI tools. It is the approach to implementation.

The Failure Pattern We See Repeatedly

In truth, most AI implementation misfires stem from neglecting one or more of the essential elements of success. We have seen companies approach us after costly mistakes, losing money because they believed technology alone could handle the heavy lifting of AI implementation.

Every failed AI implementation follows a predictable pattern: businesses put their entire emphasis on the software while overlooking the team dynamics that drive adoption. Companies often select powerful AI tools without considering employee resistance, automate processes without informing customers of changes, and pursue efficiency improvements while unintentionally overlooking the relationships that contributed to their company's growth.

The Relationship Advantage That Tech Companies Cannot Replicate

Success Story
Kaplan Early Learning

Kaplan Early Learning Company serves children from birth through age 8, with over 200 employees. They faced growing customer inquiries that strained their support team's capacity to provide personalized educational guidance to families.

Instead of hiring additional staff, they implemented ***HubSpot Breeze Customer Agent*** as part of their Service Hub package (**$450/month** for five users). The AI was trained on their educational expertise to handle routine questions while preserving human interaction for complex guidance.

The Verified Results in 6 Months:

- 25-30% of **customer inquiries were resolved** without personal intervention.
- There was a 30% **reduction in average customer service response times.**
- 37% of chat requests (about 670 chats) were processed by AI exclusively.
- Improved customer satisfaction through **faster responses.**
- Staff could focus on complex educational consultations.

According to Ben Simon, ***HubSpot Breeze*** Administrator, "Implementing Breeze customer agent this past quarter, our chat supported 37% out of 1800 chat requests without human intervention. We are continuing to build out our knowledge base to increase the self-service approach."

Their approach creates something tech-only solutions cannot replicate: **the combination of immediate AI assistance with deep educational expertise when families need complex guidance.**

The Universal Pattern of Success

Successful AI adoption across different industries follows a consistent pattern regardless of your business type:

Pattern 1: Efficiency Boost - AI handles routine tasks, allowing employees to concentrate on more valuable customer interactions and relationship development.

Pattern 2: Expertise Enhancement - Technology provides insights that make human knowledge more valuable and actionable.

Pattern 3: Service Personalization - AI allows for tailored customer experiences that strengthen personal connections.

Four Ways AI Enhances Relationships

Your relationship-based background creates **competitive advantages** that become exponentially more powerful when enhanced by **artificial intelligence.**

Advantage 1: Trust-Based Decision Making

Traditional businesses that are not slaves to stockholders make decisions based on **long-term customer relationships** over short-term profit optimization. This creates deeper **customer loyalty** and higher **lifetime value**, but sometimes limits scalability due to constraints on people's abilities.

AI Empowerment: Customer relationship management systems like ***HubSpot CRM*** and ***Salesforce Essentials*** can track relationship history, preferences, and communication patterns across unlimited customers. You maintain personal attention while scaling **trust-based decision-making** beyond the limits of human memory.

Success Story
Foothills Roofing & Exteriors

Armando Jacox Jr. and his wife Annie launched *Foothills Roofing and Exteriors* in Denver in 2018 with limited capital and no industry connections. Colorado's hailstorms created steady demand, but tracking leads, estimates, insurance claims, and payments through spreadsheets would not scale beyond a handful of jobs.

They implemented ***JobNimbus*** from day one, using the platform to automate customer communication, generate estimates with aerial roof measurements, and track jobs from first contact through final payment. What used to take hours of administrative work now happens in moments.

The results: Growth from zero to two million dollars in revenue in their first year. "Without ***JobNimbus,*** we could not do that kind of volume," Armando says. "It is impossible." Today, *Foothills Roofing and Exteriors* operates from offices in Denver and Boulder, maintains 80+ five-star Google reviews, and ranks in the top 5% of Colorado contractors. Across ***JobNimbus***' 8,000+ users, businesses report 43% average revenue growth and 24 team hours saved weekly.

Advantage 2: Local Market Expertise

Traditional businesses understand **regional preferences, seasonal patterns, and community dynamics** that national platforms cannot replicate. **AI amplifies this local knowledge into predictive intelligence.**

AI Augmentation: Location analytics tools like ***Google My Business Insights*** (free) and ***Birdeye*** can analyze local customer behavior patterns, seasonal trends, and competitive positioning to optimize your local market advantages at scale.

Success Story
Suzanne Sansom Insurance

Suzanne Sansom had been running her Katy, Texas, independent insurance agency in the Houston suburbs for over two decades. She knows her clients, understands the Gulf Coast market with its flood risks, hurricane season, the whole package. She has built real relationships in her community. But she had a problem that no amount of personal attention could solve. She could not physically touch every single renewal before clients started shopping elsewhere.

"I always felt like I was playing catch-up," *Sansom* explains. "By the time a client called upset about a rate increase, they had already gotten three quotes online."

She implemented ***EZLynx Retention Center***, an AI-powered tool that uses predictive analytics to flag at-risk renewals before clients even think about leaving. The system which runs about $350 a month, analyzes premium changes, compares policies automatically, and prioritizes which clients need a personal call versus which ones just need a reassuring email.

The results after three months:

- **Retention** rate jumped from the industry average of 75% to over 90%.
- Cross-selling revenue increased as the system identified **coverage gaps**.
- Her team now touches 100% of renewals (previously impossible).
- Each renewal **saves approximately 15 minutes of manual comparison wor**k.

"For the first time in 22 years, we have a system and workflow," *Sansom* says. "We can now cross-sell, round our accounts, and improve our retention and revenue. There is honestly nothing like it. We touch every single person, every single time."

What makes this work: The AI handles the data crunching and prioritization. *Suzanne* and her team handle the conversations. Her local expertise of knowing which clients need flood coverage updates before hurricane season, which neighborhoods are seeing rate increases, and which commercial clients are expanding becomes more valuable, not less. The technology tells her who to call. Her two decades of Gulf Coast experience tell her what to say.

Advantage 3: Flexible Service Customization

Relationship-based businesses excel at adapting services to individual customer needs, but human limitations restrict scalability. **AI systems can personalize service delivery at scale while maintaining flexibility**.

Success Story
Trio Heating and Air

David and *Michael Katz* had no experience in the HVAC industry when they founded *Trio Heating and Air* in San Jose in 2021. What they did have was a willingness to learn and a commitment to doing right by their employees and customers. Their first year brought in $300,000—enough to survive, but nothing special. Two years later, they hit $10 million in annual revenue.

The *Katz brothers* credit much of their growth to implementing ***ServiceTitan***, an AI field service management platform that gave them visibility into every aspect of their operation. But here is what makes their story worth telling: they did not let the software replace their values. They let it amplify them.

"We told everyone upfront, 'We are not going to take anything from this business before we know that everyone is getting paid really well and they are well taken care of,'" David explains. "And once everyone is happy, then it'll be our turn. Not before."

ServiceTitan, which averages $200 a month per technician, gave them the data to track technician performance, optimize scheduling, and identify which marketing campaigns actually drove revenue. But the Katz brothers used those insights to support their team, not squeeze them. They could see which technicians needed coaching, which customers were at risk of leaving, and where to focus their limited time.

The results:

- Revenue grew from $300,000 to $10 million in two years (3,233% growth).
- Named "Fastest-Growing HVAC Company in the Nation" by ***ServiceTitan.***
- Earned a spot on the **Inc. 5000 list**.
- Won 4 out of 5 national awards from Bryant.
- Grew from 51-200 employees.
- Planted over 10,000 trees through their "one tree per job" initiative.

What makes this work: The software handles scheduling, dispatching, customer history, and performance tracking. David and Michael handle the culture. Their local expertise—understanding the San Jose market, knowing which neighborhoods need what services, building relationships with Bay Area homeowners—becomes the foundation that technology builds upon, not something technology replaces.

Advantage 4: Employee-Customer Relationships

Traditional businesses benefit from employees **who know customers personally** and understand their preferences. **Human memory and capacity limitations restrict this advantage. AI amplifies personal relationships on a scale.**

Success Story
Paumanok Veterinary Hospital

Dr. David Hensen and *Dr. Deirdre "Dee" Hensen* have run *Paumanok Veterinary Hospital* in Patchogue, New York, since 1992. The husband-and-wife team built their Long Island practice on personal relationships. Dr. Dave speaks fluent Spanish to serve the local community, Dr. Dee specializes in birds, rabbits, and guinea pigs that other clinics turn away. With five veterinarians and deep roots in the community (they have raised over $68,000 for local causes), their reputation for compassionate care kept families coming back for generations.

But their outdated server-based software was holding them back. The system had not been updated in years, tech support was nearly nonexistent, and the clunky interface forced the team to keep paper records alongside their digital ones. **Confusion and inefficiency piled up.** Veterinarians who wanted to check on patients from home could not access files remotely. Worst of all, doctors spent hours after their shifts finishing medical documentation instead of going home to their families.

"This outdated approach resulted in confusion and inefficiencies," recalls Brian Bernatzky, Hospital Manager. "Especially as doctors increasingly sought the flexibility to work remotely."

They made the switch to ***Digitail***, a cloud-based practice management system with built-in AI capabilities called ***Tails AI.*** The AI listens during patient exams, captures the conversation, and transforms it into properly formatted medical notes—no typing, no dictation after hours, no deciphering handwritten scribbles.

First Year Results:

- **8 minutes saved per patient note** using AI-powered documentation.
- **10+ hours reclaimed per week per veterinarian**—multiplied across five doctors, that is **50+ hours weekly** the team gets back.
- **99% accuracy** on AI-generated notes, requiring only quick review.
- **More comprehensive records** than manual note-taking ever produced.
- **No more late nights** finishing paperwork after a full day of appointments.

"We were all amazed by the results," *Bernatzky* says. "***Tails AI*** flawlessly transcribed and transformed the conversation into the SOAP format, extracting only the pertinent information. It was akin to magic. We did not realize it would be such a game-changer for the way we operate."

The AI does not make medical decisions. It captures what happens in the exam room so veterinarians can focus on the animal in front of them instead of splitting attention between patient and keyboard. Dr. Dave still brings his 40+ years of diagnostic expertise to every case. Dr. Dee still knows which guinea pig owners need extra reassurance. The technology handles the documentation; the doctors handle the medicine.

"AI is making our jobs easier without taking away the control over medical decisions," *Bernatzky* explains. "It is like having a personal assistant who you can bounce ideas off of, get decision support, and reassurance in your presumptive diagnoses."

The Competitive Reality Check

Tech-native companies excel at attracting customers with innovation and promises of efficiency, but struggle with **customer retention** because they cannot replicate the relationships that drive long-term loyalty.

The evidence shows: **customers will initially try automated services for convenience, but they often return to relationship-based businesses for trust, expertise, and personalized attention** when problems arise or their needs become more complex.

Your AI-enhanced relationship approach **creates competitive advantages** that become stronger over time.

The betrayal feels personal. These employees are real people in your life, not just names on payroll. They are the ones who stayed late during tough times. Who brought casseroles when your mom was sick, and who trusted you when nobody else would give you a chance.

Now you are googling "AI automation tools" and wondering which of their jobs you are researching away. One owner told us, "I felt like Judas every time I looked at AI software. These people built my business. How could I even think about replacing them?"

Here is what that owner discovered: AI did not replace anyone. It gave his team superpowers.

The Competition You Cannot See Coming

The conversation that changed everything happened over coffee. *Lisa*, who runs an accounting firm, looked exhausted. "I do not understand what is happening," she said. "My clients keep mentioning this new service that does their books for half my rate. But how can online-only companies provide the same level of service?"

We had to break the hard truth: her competitor was **providing better service for less money using AI tools** that improved efficiency and accuracy for accounting work.

This conversation happens weekly, with business owners discovering they are being outmaneuvered by competitors they did not even know existed last year. The harsh truth is that while you are debating whether to adopt AI, other companies are already using it to capture your customers.

The Strategic Advantage of Moving Now

The AI opportunity has an expiration date, and that expiration date is **approaching faster than you realize**.

Currently, smart business owners consistently outperform their competitors by **leveraging AI tools** that showcase the best of everything they have built. They win customers with **personalized service delivered at technological speed**. They solve problems with **expertise enhanced by data insights**. They create experiences that can combat decades of relationship-building with cutting-edge efficiency.

What is concerning is that while you evaluate AI options, competitors are gaining market share every day. The window for competitive transformation is wide open today, but it narrows as more businesses successfully adopt AI, while your business waits for ideal conditions that may never come.

Every thoughtful step you take **builds advantages that are impossible for others to copy**, but only if you take those steps before your market becomes saturated with AI-enhanced competitors.

This is not fearmongering. This is pattern recognition based on consulting with businesses across various industries over the past few years. The ones that moved early captured market share. The ones that waited lost customers to competitors they never saw coming.

The technology is ready. The proven step-by-step method is in place. **The opportunity exists today.** The only question is whether you will seize the competitive edge that could shape your business future. Or will you wait and watch other firms capture the customers you are losing while you refine your AI evaluation process?

Chapter 3

Three Pillars of AI Protocols

The **Three Pillar AI Success Strategy** promotes **people over technology** and focuses on getting your team on board with implementing **AI systems**.

Take note of the **$50,000 mistake** you are about to make. That is the average amount **small and medium-sized businesses waste on their first AI implementation**. Not because **technology fails**. Because they bet everything on tools while **ignoring the people who must use them**.

Picture that you have just signed a contract for **AI software** that costs more than your first car. The demo was flawless. The **ROI projections** made your accountant smile. Six months later, the software is collecting digital dust while your team finds creative ways to work around it.

Your family's security depends on getting this right. Your employees' mortgages depend on your business thriving. Clearly, there is no pressure here.

Instead of depending solely on AI tools, this approach highlights three pillars of successful AI implementation. Every effective AI deployment relies on this groundwork:

1) Technology Systems

2) Team Development

3) Customer Engagement

Each pillar supports the others, creating a strategy that maintains **trusted relationships** while building **technological skills**. **Removing one component can reduce the overall stability of an AI project.** Establishing a foundation provides a framework that supports company growth and facilitates critical relationships with clients, employees, and vendors.

Why Many AI Implementations Fail

In truth, most **AI implementation failures** stem from neglecting one or more of the essential pillars of success. We have seen companies approach us after costly mistakes, losing money because they believed the tool on its own could handle the heavy lifting of AI implementation.

AI Realities

- **67% of AI projects fail because of pushback from the team.**
- **34% of customers complained about reduced personal service.**
- **28% of projects fail because the technology does not integrate with existing processes.**
- **45% of AI implementations collapse because business owners chose tools based on vendor hype, not business needs.**

Often, **technology is not the big problem**; rather, it is people who present the real challenge.

AI Failure Stories Are Not a Reason to Avoid Adoption

Just because there are stories of companies failing at AI does not mean you should avoid implementing AI in your company. While some organizations have struggled with adoption due to employee resistance, poor integration, or misguided tool selection, these challenges highlight the need for a **thoughtful approach**, not a reason to abandon progress.

We share these stories so you can learn from these experiences and focus on strategies that prioritize **people, communication, and coordination with your business needs**. By recognizing common pitfalls and proactively addressing them, your company can successfully leverage AI to drive profitable results and avoid the expensive errors others have made.

What Vendors Do Not Tell You About Implementation

Vendors sell you technology. They do not prepare your team or communicate with your customers. That is why 67% of implementations fail even when the technology works perfectly.

When vendors promise "seamless integration," they mean that software can connect with your systems. They do not mean your employees will use it effectively or your customers will appreciate the changes. Although business owners are responsible for the reactions of their employees and customers, they often overlook the importance of including them in discussions about AI integration before signing vendor contracts.

Other Secrets AI Vendors Keep

Most AI vendors will show you impressive demos with perfect data and ideal conditions. **Ask the vendor what happens when your internet connection fails,** when employees resist using the system, or when customers complain about automated responses. **Vendors who cannot answer these questions are selling dreams, not solutions.**

When using lower-cost vendor solutions without live sales representatives, it is necessary to depend on online reviews of vendor products.

We have watched companies waste thousands of dollars on AI tools that promised "revolutionary results" but delivered improvements of only 5% at best. Vendors who offer substantial assurances without providing clear explanations or evidence warrant thorough due diligence.

The Three Pillar Overview

Pillar One: Technology Systems

Focus on choosing and utilizing AI tools that enhance, rather than replace, human skills. Instead of choosing systems based on their maximum ability to automate processes, evaluate AI solutions based on their ability to strengthen customer relationships and support your team's expertise.

Key Questions for Technology Selection:

1. Does this tool integrate smoothly with our existing systems?
2. Will this enhance our team's capabilities or bypass them?
3. Can we implement this without disrupting customer relationships?
4. What happens when the technology fails or needs maintenance?
5. Is the vendor providing realistic timelines and support?

Pillar Two: Team Development

Position AI as a career enhancement, not a job replacement. Through ongoing training and communication, your employees will become AI-powered experts, not victims of technological change. This pillar cultivates internal AI supporters within your company who understand how AI enhances their skills.

Key Elements of Team Development:

- Early involvement in AI tool selection.
- Comprehensive training before implementation.
- Clear communication about how AI improves their roles.
- Recognition and rewards for successful adoption.
- Ongoing support during transition periods.

Success Story Insurance Agency

Susan runs an insurance agency in Washington, serving 185 clients. Her team worried that AI would eliminate their jobs, which created resistance to AI technology improvements.

She implemented ***Applied Epic AI*** ($298/month) using a team-inclusive approach. Employees participated in AI tool selection, received comprehensive training, and were positioned as "AI-Enhanced Insurance Specialists."

Results in 8 months:

- The agency was 48% faster at policy research.
- 73% improvement in quote accuracy.
- 35% reduction in claim processing time.
- 97% employee satisfaction.
- Team members received promotions and salary increases based on improved productivity.

Susan's success came from addressing employee concerns before they became resistance problems.

Pillar Three: Customer Engagement

Maintain transparency and trust throughout your AI implementation. Instead of making changes that surprise or concern customers, this pillar rewards ongoing communication that positions AI development as an improvement to the personal service your customers already value.

Key Elements of Customer Engagement:

- Pre-implementation communication about upcoming improvements.
- Clear explanation of how AI enhances personal service.
- Easy access to human support when needed.
- Ongoing feedback collection and response.
- Transparency about what AI handles versus human expertise.

The Coordination Approach - All Three Pillars Working Together

The Three Pillar Protocol is most effective when its elements work together, each strengthening the others to drive faster AI implementation success. This integrated approach ensures that potential gaps are addressed proactively and resources are allocated efficiently. By leveraging the strengths of each pillar, organizations can create a strong framework that promotes innovation and maximizes outcomes.

Success Story
Badger CPA

Badger CPA in San Antonio, TX, is led by Shannon Badger, who understands that AI implementation success depends on the Three Pillars.

The company transformed their traditional accounting practice using AI while **strengthening client relationships.** Here is their step-by-step approach, exactly what they did, and how much it cost:|

> Shannon realized that **successful AI implementation** required addressing technology, team capabilities, and client expectations simultaneously.

Timeline for Implementation:

Months 1-2: Shannon trained the team on ***Receipt Bank*** and ***QuickBooks Advanced***, focusing on AI as a way to improve job efficiency, rather than replacing jobs. Shannon made it clear that faster processing of expenses meant more time for client strategy sessions.

Month 3: The firm began a client education campaign highlighting how AI's improved accuracy would lead to better financial insights and faster reporting. Clients received personal calls from the team explaining the AI updates.

Months 4-6: The *Badger CPA* system was rolled out systematically across all clients, continually monitoring relationships and gathering customer satisfaction feedback.

Results After Nine Months:

- **Processing time** for monthly bookkeeping decreased from eight hours to 2 hours per client.
- **Client satisfaction increased** by 67% through improved accuracy and faster reporting.
- **Revenue grew 34%** per client through value-added advisory services, which were enabled by gains in efficiency.
- Employee **job satisfaction improved** because of a new focus on client relationships and the end of repetitive, time-consuming data entry.

- **Client retention reached 98%** due to enhanced service quality and an increase in personal customer satisfaction.

What distinguishes *Badger CPA* from less successful AI projects is that **they tackled employee concerns early**, preventing staff resistance. The team communicated with clients before implementing changes from AI technology. Most businesses approach these implementations in exactly the opposite manner. Shannon did not have a magic tool. She had the right pillars.

The Coordination Approach That Creates Success

The power of the Three Pillar Protocol comes from coordinated implementation. Each pillar reinforces the other, creating momentum that accelerates the success of AI implementation.

Weeks 1-2: Simultaneous Groundwork

Pillar One: Research AI tools focusing on integration and team adoption.

Pillar Two: Begin employee education about AI benefits and career development.

Pillar Three: Develop customer communication emphasizing service enhancements.

Weeks 3-4: Coordinated Preparation

Pillar One: Configure AI tools in a test environment with a limited scope – avoid mass integration.

Pillar Two: Provide hands-on training using real business scenarios.

Pillar Three: Send pre-implementation communication to clients about upcoming improvements.

Weeks 5-6: Synchronized Launch

Pillar One: Deploy AI tools with monitoring and optimization.

Pillar Two: Support employees during transition with ongoing training and coaching.

Pillar Three: Communicate launch, emphasizing immediate customer benefits.

Weeks 7-12: Integrated Optimization

Pillar One: Expand AI capabilities based on employee and customer feedback.

Pillar Two: Team develops advanced skills and internal expertise.

Pillar Three: Share results with customers and gather their feedback for improvement.

The Strategic Baseline for Competitive Excellence

By using the Three Pillar approach, you can **build lasting competitive advantages** that snowball over time, not just offering short-term benefits that others can easily match.

Relationship-Enhanced AI: Your methodical approach preserves and enhances personal connections while gaining technological capabilities - creating advantages that **tech-only competitors cannot match.**

Employee Excellence: Success across team development **creates loyalty and AI expertise** that supports business growth and market expansion.

Customer Leadership: AI implementation strengthens and does not threaten relationships, creating a market position that national competitors cannot replicate.

The Three Pillar Implementation Protocol Shifts

AI adoption has progressed from a risky project to a means of enhancing employee and customer relationships, as well as improving your competitive advantage. The next chapter will show you exactly how to choose AI vendors and negotiate contracts that protect your investment while ensuring successful implementation.

AI adoption is about evolving away from seeing artificial intelligence as just a risky project experiment to becoming a powerful means of enhancing employee and customer relationships. This approach improves your competitive advantage.

Building on your new perspective, the next chapter will guide you through selecting AI vendors and negotiating contracts that safeguard your investment and set the stage for successful implementation.

Chapter 4

AI Readiness

This chapter outlines **practical strategies in project planning, team collaboration, and technology selection** to help you launch successful AI solutions right from the start. By focusing on these key areas, you will be equipped with steps and clear guidance for **smooth and successful AI implementation**.

Jennifer's daughter asked at dinner, "Mom, are we going to be okay?"

The question hit like a punch. Jennifer had been **researching AI tools all week**, calculator in hand, trying to figure out if she could afford the investment in AI for her company. Her daughter had overheard worried phone calls about competitors using AI.

"That's when it hit me," Jennifer said. "This wasn't just about my business. My daughter's college fund, our mortgage, my parents' retirement help, everything depended on choosing AI tools wisely."

If you are feeling that pressure, you are not alone. Every business owner we work with carries the same burden of investing in something that they do not fully understand when their family's future hangs in the balance.

Most businesses rush into AI selection without assessing whether their organization is ready. This chapter is geared to prevent avoidable missteps by helping you build the internal readiness that separates successful implementations from expensive failures.

Assessing Your AI Readiness

Team Confidence Assessment

Your team's attitude toward AI will determine whether your implementation succeeds or fails. Before selecting any technology, you need to understand your employees' comfort zones, fears, and expectations. Some folks will see AI as a tool that saves time and eliminates the drudgery of repetitive tasks; others will see it as a threat waiting to automate them out of a job.

You cannot build trust on top of silent resistance. Talk openly about what AI will and will not change in your company and with team members' jobs. The smartest companies invest first in mindset—teaching their teams that AI does not replace individual judgment; it amplifies it. Once people see AI as an assistant, not an adversary, technology decisions become much easier.

Questions For Your Team:

1. **How comfortable are you with learning and adapting to new technology?**

2. **What concerns or hesitations do you have about using AI in our business?**

3. **In what ways do you think AI could make your work easier or more effective?**

4. **What would make you genuinely interested or excited to try AI tools?**

5. **What might make you hesitant or resistant to using them?**

Create anonymous surveys to get candid replies. The results will guide your communication strategy and training approach. You will find a sample employee survey in the appendix of this book.

Technology Infrastructure Evaluation

AI tools require reliable infrastructure. If your internet drops, your data is disorganized, or your systems do not communicate with each other, the smartest AI in the world will not save you. Think of it like hiring a world-class chef but giving them a broken stove and no ingredients. The outcome will still be a mess. Before rolling out any AI system, make sure your systems are solid: stable internet, clean data, compatible software, and clear security protocols. Without those basics, every automation or insight will be built on quicksand.

To make sure your systems can handle modern AI tools, review these minimum standards:

1. **The CPU Processor.** Use at least an **Intel Core i5**, **AMD Ryzen 5**, or **Apple M1** processor. These can handle most AI-assisted tasks, from chat interfaces to data dashboards. For heavier workloads such as data analysis, image generation, or large spreadsheets, upgrade to **Intel i7**, **Ryzen 7**, or **Apple M2/M3**.

2. **RAM Memory. 16 GB RAM** keeps AI tools running smoothly, especially when multiple browser tabs or apps are open. Systems with only 8 GB can function, but will lag under real-world use.

3. **Storage.** Install a **512 GB solid-state drive (SSD)** or larger. SSDs read and write data far faster than older hard drives, reducing loading times and improving reliability. If your business works with video, large images, or big datasets, expand to 1 TB or use external SSD backup drives.

4. **Operating Systems.** Stay current with **Windows 11 Pro** or **macOS Monterey or later**. These provide better security and compatibility with the newest AI programs. Older operating systems can block updates or integrations that AI platforms rely on.

5. **Graphics (GPU).** For basic business tasks, integrated graphics are fine. For visual work—such as design, video, or data visualization—a **dedicated GPU** like an NVIDIA GTX 1660 or better offers smoother performance.

6. **Internet Connection.** AI tools depend on constant, stable access to the cloud. Use **high-speed broadband** with at least **25 Mbps download and 10 Mbps upload per user**. Unstable Wi-Fi is the silent killer of AI productivity.

7. **Security and Backup.** Use **business-grade antivirus software**, enable **multi-factor authentication**, and perform **daily automatic cloud backups** through services like Google Drive, OneDrive, or Dropbox. Protecting your data is not optional—it is the backbone of AI reliability.

Critical Infrastructure Questions:

1. Is your internet connection reliable enough for cloud-based AI?
2. Are your current software systems compatible with AI - integration?
3. Do you have adequate data security measures in place?
4. Is your customer data organized and accessible?
5. Can your current hardware support new AI applications?

If you answer "no" to more than 2 questions, address infrastructure gaps before implementing AI.

Customer Relationship Strength Check

Strong customer relationships are the baseline of every successful AI improvement. When trust runs deep, technology can enhance it. When relationships are already shaky, automation only makes the cracks show faster.

Signs Your Customer Relationships Are Strong:

☐ You keep more than **80%** of your customers year after year.

☐ You communicate regularly—not just when there is a bill or a problem.

☐ You know personal details about what your customers value or prefer.

☐ Customers are comfortable giving you honest feedback.

☐ Referrals come in steadily from people who have already worked with you.

If these indicators are weak, fix the human side first. **Do not hand off fragile relationships to automation and expect AI to repair them.** Strengthen connections and trust, then bring in AI to extend, not replace, what already works.

AI Budget Estimates by Business Size

AI costs vary significantly based on the business size and complexity. Here are budget ranges for your planning purposes:

Micro Business (1-5 employees):

- Entry-level AI tools: $50-$200/month
- Mid-tier solutions: $200-$500/month
- Advanced implementations: $500-$1,000/month

Small Business (6-25 employees):

- Entry-level AI tools: $200-$500/month
- Mid-tier solutions: $500-$1,500/month
- Advanced implementations: $1,500-$3,000/month

Growing Business (26-50 employees):

- Entry-level AI tools: $500-$1,000/month
- Mid-tier solutions: $1,500-$3,500/month
- Advanced implementations: $3,500-$7,000/month

Realistic ROI Timeframes: Expect 3 to 6 Months

AI may not yield immediate results, but it can provide a return on investment faster than many other business improvements. Most companies start seeing measurable returns within three to six months—once the team adjusts, data flows correctly, and workflows stabilize. Expect some trial and error early on. Quick wins can happen, but lasting results come from steady use, not instant transformation. Think of AI as a new employee who needs a few months to learn the ropes before showing full value.

Months 1-2: Implementation Phase

- ✓ Expect costs without immediate returns.
- ✓ Focus on team training and system setup.
- ✓ Measure progress, not profit.

Months 3-4: Transition Phase

- ✓ Begin seeing efficiency improvements.
- ✓ Some revenue impact starts appearing.
- ✓ Continue refining processes.

Months 5-6: Results Phase

- ✓ Measurable ROI becomes apparent.
- ✓ Team is fully comfortable with systems.
- ✓ Customer satisfaction metrics improve.

Months 7-12: Optimization Phase

- ✓ ROI accelerates as efficiency compounds.
- ✓ Expand AI capabilities based on success.
- ✓ Scale implementations across more areas.

Hidden Costs to Anticipate

AI tools rarely cost only the monthly subscription price. The fine print often lives outside the invoice—buried in setup, training, and maintenance. Budgeting for these extras upfront keeps you from being surprised later.

Common Hidden Costs:

- **Integration with your current systems:** $500–$5,000, depending on complexity.
- **Employee training and adjustment time:** 10–20 hours per person before full productivity returns.
- **Data cleanup and organization:** $1,000–$10,000 to make your information usable for AI.
- **Additional software licenses:** $50–$500 per month for connected tools and plug-ins.
- **Ongoing technical support:** $100–$1,000 per month for troubleshooting and updates.
- **System customization:** $1,000–$15,000 to make the tool fit your actual workflows.
- **Backup and security upgrades:** $100–$500 per month to protect new data pipelines.

The smartest AI budgets include both the price of the tool and the cost of making it work well. AI is not a one-time purchase; it is a system you build, refine, and maintain over time.

Creating Your Communication Strategy

The way you communicate about artificial intelligence greatly influences the **successful adoption of your strategy**, both within and outside your company. Employees need to feel included, not replaced. Customers need to see improvement, not distance. **Clear, honest communication** builds confidence on both sides.

Businesses that combine **AI efficiency with relationship prowess** do not just compete better—they dominate their markets because they offer something nobody else can match. Those companies took part in conversations with everyone involved with their company. You will have talks with your **team, competitors, customers, and your community** about AI. We created opening lines to position yourself and create an **open dialogue**.

Internal Messaging to Employees

Your team's attitude toward AI will decide whether the rollout succeeds or fails. Communicate early, involve them in decisions, and address concerns head-on. A transparent approach turns potential skeptics into supporters.

Week 1: Initial Announcement

- ✓ Explain that your business is exploring **AI solutions**.
- ✓ Emphasize that **employee input** will guide the process.
- ✓ Talk openly about job security and the value of **human skills**.
- ✓ Highlight opportunities for learning and your team's **career growth**.

Weeks 2–3: Education Phase

- ✓ Clarify what AI can and cannot do in your specific business.
- ✓ Share real **success stories** from your industry.
- ✓ Connect benefits directly to their **daily work**.
- ✓ Keep discussions open and encourage **honest questions**.

Week 4: Decision Involvement

- ✓ Invite employees to help evaluate tools or test **demo versions.**
- ✓ Ask for feedback on what feels useful or frustrating.
- ✓ Use their insights to shape your **final choice.**
- ✓ Reinforce that this is a **shared investment** in making work easier.

Weeks 5–6: Preparation and Launch

- ✓ Announce the final **tool selection and rollout plan**.
- ✓ Share the **timeline, training dates, and success benchmar**ks.
- ✓ Explain how **performance** and progress will be tracked.
- ✓ Celebrate the **launch of your AI program** together with employees.

External Messaging to Customers

Customers need to know that AI will make their experience better—not colder. The goal is to show that technology improves service while keeping **personal connection** at the center.

Pre-Implementation (2–3 Weeks Before Launch)

- ✓ Announce upcoming **improvements** in how you will serve your customers.
- ✓ Explain how AI will enhance **accuracy, speed, or convenience**.
- ✓ Reassure folks that **human help** remains fully available.
- ✓ Invite **feedback** or questions through your normal contact methods.

Implementation Week

- ✓ Clearly outline what has changed in your business and why.
- ✓ Point out the **immediate customer benefits**, like faster responses or improved access.
- ✓ Keep **human support** easy to reach.
- ✓ Thank customers for their patience during the **adjustment period**.

Post-Implementation (Ongoing)

- ✓ Share **measurable results** and customer wins.
- ✓ Ask for **feedback** about their experience.
- ✓ Adjust based on what you hear.
- ✓ Publicly recognize **progress** and thank customers for growing with you.

Handling Resistance and Concerns

Every major change brings questions, and AI is no exception. Both employees and customers will have **legitimate worries**. The key is to address **them early, honestly, and often.** Silence creates rumors; transparency builds trust.

Common Employee Concerns and How to Respond

Concern: "AI will replace my job."

> **Response:** AI is here to handle **repetitive, low-value tasks** so you can focus on the work that requires your **judgment, creativity, and expertis**e. We are using AI to **grow the business**, not shrink it. Growth means more opportunities for skilled people, not fewer.

Concern: "I am not good with technology."

> **Response:** You do not need to be. We will provide **step-by-step training** and **ongoing support**. The tools we are choosing are built for **regular users**, not IT specialists. Our goal is to **simplify your workload**.

Concern: "This seems complicated and time-consuming."

> **Response:** Every new system takes a little time to learn, but the payoff is worth it. Once you are comfortable, AI **will save hours each week** and remove much of the repetitive work that slows you down today. We will **roll it out gradually** so you can adjust at your own pace.

Common Customer Concerns and Responses

Concern: "I do not want to talk to a robot."

> **Response:** You will still work with the **same people you know and trust**. AI simply handles quick or repetitive questions so our team can spend more time on what matters most—**personal attention and meaningful service**.

Concern: "Automated systems never work right."

> **Response:** We understand that frustration. That is why we are **testing and training** our AI systems carefully before launch. The goal is not to replace people but to help our team **respond faster** and **serve you better**.

Concern: "My information will not be secure."

> **Response:** Your **privacy** is a top priority. We work only with AI vendors who meet **strict data protection standards**, and we have added extra layers of **security**. In most cases, your information is safer with these systems than with manual processes.

Setting Success Metrics

You cannot manage what you do not measure. Clear, simple **metrics** tell you whether your AI investment is working—or just creating noise. Track results in four areas:

efficiency, **relationships**, **revenue**, and **team satisfaction**. Together, these measures give you a complete picture of progress. *See chapter 12 for more on metrics*.

Efficiency Gains

AI should make work **faster** and more **accurate**, not just flashier. Focus on measurable time savings and fewer mistakes.

Key Efficiency Metrics:

- **Hours saved** on repetitive tasks each week.
- Reduction in **data-entry or processing errors**.
- **Faster response times** for customer inquiries.
- **Shorter turnaround** for transactions or services.
- Fewer manual **follow-ups** or corrections required.

Relationship Quality Measures

AI success is not only about numbers—it is about maintaining trust. The goal is to strengthen relationships, not automate them away.

Customer Relationship Metrics:

- **Customer satisfaction** or feedback scores.
- **Net Promoter Score (NPS)** trends.
- **Customer retention** and repeat business rates.
- **Referral frequency** from existing customers.
- Complaint frequency and **resolution time**.

Employee Relationship Metrics:

- **Employee satisfaction** with AI tools and workflows.
- **Voluntary turnover** trends.
- **Internal promotions** and advancement rates.
- Measured **productivity** or project completion speed.
- Quality of **team collaboration** and communication.

Revenue Impact Indicators

AI should eventually show up on your **income statement**. The strongest indicators reveal both **short-term gains** and **long-term growth**. (*See Chapter 12 for more on Metrics.*)

Revenue Metrics:

- **Sales growth rate** over time.
- **Average transaction** or deal size.
- **Customer lifetime value**.
- **Cost to acquire** each new customer.
- Success rate of **cross-sells and upsells**.
- **Revenue generated per employee**.

Employee Satisfaction Tracking

AI adoption succeeds only when your team stays **engaged** and **confident**. Monitor how your employees **feel** about the change, not just how fast they adapt.

Team Satisfaction Metrics:

- Employee surveys on **comfort and confidence** using AI.
- **Training completion rates** and skill improvement levels.
- Voluntary daily or weekly **use of AI tools**.
- Suggestions for **tool improvements** or workflow changes.
- Development of internal **AI champions** or mentors.
- Increases in requests for **growth or career training**.

Your Immediate Action Plan

Week 1: Trust Baseline: Complete **Team Confidence Assessment.** Begin transparent communication with your team and key customers about your AI exploration and commitment to preserving what they value most about your business.

Week 2: Technology Research: Research AI tools specific to your industry. Prioritize solutions that **enhance human capabilities** rather than replace them, ensuring they integrate smoothly with your current systems.

Week 3: Implementation Planning: Design your **three-pillar preparation strategy** covering **technology selection, employee training,** and **customer communication.** Establish success metrics that measure both efficiency gains and relationship quality.

Week 4: Strategic Commitment Make the definitive decision to begin AI implementation using the **TRACK Framework,** which is outlined in Chapter Six. Set **your 9-month transformation timeline** and communicate your commitment to a systematic, relationship-preserving approach to **AI adoption**.

Stay on Task

It is completely normal to be concerned that AI implementation is a complex undertaking, which may leave you feeling discouraged. Keep in mind that while you read this guide, three of your **competitors** are implementing AI systems that will capture your customers six months from now. The question is not whether AI will change your industry; it is whether you will **lead that change** or become its victim.

Chapter 5

AI Vendor Decision Framework

*Mike** will never forget the vendor's promise of "**Seamless integration with zero downtime.**"

Eight weeks later, his HVAC business was hemorrhaging **$3,000 a day**. The "seamless" integration had crashed his **scheduling system** during Phoenix's hottest June on record. Customers screamed. Technicians sat idle. The vendor's support line rang busy.

"I trusted them," Mike said. "They had testimonials, awards, and a beautiful website. What they were missing was any clue how our business actually worked."

By the time Mike extracted his company from the mess, he had lost **$47,000** and **six employees**. The vendor? They offered a partial refund and wished him luck.

When evaluating AI vendors that claim to revolutionize your business, **the vendor with the most impressive demo often ends up causing the most issues**. It is easy to be wowed by AI tools. Then, months later, their "perfect" solution does not integrate with existing systems, requires a computer science degree to run, or harms your valued customer relationships.

This **decision framework** is designed to minimize costly errors, providing targeted questions and industry-*specific examples to guide your evaluation process effectively.*

Choosing the Right AI Tools

AI tools are simply **applications**—software programs built to help your business think and work smarter. They are not mysterious or reserved for tech giants. Most are designed with your **specific industry** in mind, whether you run a restaurant, manage real estate, or provide professional services. The key is to look for tools that **solve real problems** you already face, not shiny gadgets that promise magic.

A good AI tool feels like hiring an extra set of hands that never gets tired, forgets details, or calls in sick. Before buying anything, make sure the tool **integrates with what you already use** and that your team can actually understand it. AI should **simplify your workday**, not complicate it.

Research Approach for Your Industry

Every industry moves at its own pace in adopting AI. Some are already deeply invested in automation and analytics, while others are just testing the waters. Your research should match **where your industry is**—not where the headlines say it is. The goal is to learn from people already using AI successfully, not to reinvent the wheel.

Practical Research Steps:

1. **Study the front-runners.** Look for businesses like yours that have made AI work in the real world. Read their case studies, note which tools they used, and what lessons they share.
2. **Join the conversations.** Industry forums, LinkedIn groups, and trade associations often discuss which AI tools actually deliver results. Listening here can save you costly trial and error.
3. **Show up where it matters.** Attend conferences, webinars, and expos focused on your line of work. The best insights often come from casual hallway conversations or Q&A sessions, not sales brochures.
4. **Read what professionals read.** Trade journals, newsletters, and respected websites often feature side-by-side comparisons and honest reviews of AI tools built for your business type.

The smartest research does not chase hype. It looks for proof and evidence of what works for people doing the same kind of work you do.

Questions for AI Application Vendors

Never accept **vendor promises at face value**. Ask tough questions that reveal implementation realities:

Integration:

1. What specific systems does your **AI integrate** with?
2. How long does typical integration take for **businesses of our size**?
3. What **technical support** do you provide during integration?
4. What happens when system connection issues arise?

Implementation:

5. What **training** do you provide for our team?
6. What is the typical **timeline** from purchase to full implementation?
7. How many employees will need **training**?
8. What **ongoing support** is included in the pricing?

Performance:

9. What **measurable results** do similar businesses achieve?
10. Can you provide **references** from businesses in our industry?
11. What happens if the AI does **not deliver promised results**?
12. How do you **measure success** for customers?

Reliability:

13. What is your system **uptime percentage**?
14. What happens when your **system goes down**?
15. How quickly do you respond to **technical issues**?
16. What **backup systems** do you recommend?

Red Flags in Vendor Promises

AI vendors make big promises. Your job is to **separate confidence from sales talk**. When a company oversells, it usually shows early—if you know what to look for. Pay attention to the **details they skip**, not just the claims they make.

Significant Warning Indicators:

- Claims of **"revolutionary results"** without showing real numbers or examples.
- No **references from businesses similar to yours**.
- Evasive or unclear answers about **setup or integration timelines**.
- No explanation of **team training requirements**.
- High-pressure tactics to **sign a contract before you have time to think**.
- Buzzwords like **"seamless" or "plug and play"** instead of honest timelines.
- No discussion of **risks, limitations, or potential problems**.
- Refusal to offer a **trial period or pilot program**.

If a vendor cannot be **transparent before you pay them**, imagine how they will respond after you do. Select partners who **speak plainly, share proof, and respect your need to understand** exactly what you are buying.

How to Read Reviews and Case Studies

Not every review or case study tells the full story. Many are written to sell, not to inform. Your goal is to spot the difference between **polished marketing and real experience**. Approach every glowing success story with curiosity and **healthy skepticism**.

When Reading Online Reviews:

- Look for **specific examples** of how the tool was used, what worked, and what did not.
- Focus on **recent feedback**, ideally within the last six months. AI tools evolve fast.
- Watch for **consistent patterns** across multiple users rather than single opinions.
- Be cautious of reviews that are **all positive or all negative**; balance usually signals truth.
- Compare reviews on **more than one platform** to confirm credibility.
- Give extra weight to reviews from **businesses similar in size and industry** to your own.

When Evaluating Case Studies:

- Confirm the company featured is **comparable to yours** in scale and market.
- Look for **specific metrics**—numbers that prove results, not vague claims of "improvement."
- Pay attention to **challenges mentioned**; honest case studies show the bumps, not just the finish line.
- Ask if the **timeline** seems realistic or too good to be true.
- Note any mention of **ongoing costs or support** after launch.
- Check whether the **business still uses the tool** or quietly moved on.

Strong reviews and case studies reveal the **process, the struggles, and the real results**. Those are the stories worth trusting.

The Three-Pillar Evaluation Framework

Before you watch a single AI demo, clearly **define what automation success means for your business**. Compare vendors based on **established criteria, rather than being impressed by features** during demonstrations.

Pillar 1: Business Fit (Most Important)

- Does this AI application **solve our actual problems** or create new ones?
- Will this initiative **strengthen our relationships with customers**, or introduce additional complexities?
- Is it feasible for our team to **acquire the necessary skills and utilize this AI system efficiently**?

Pillar 2: Technical Integration

- Is this AI software **compatible with our existing systems**, or would it require significant and costly customization?
- Can we implement this system **without disrupting existing workflows**?
- Are the **data security and privacy protections** adequate? Who owns the data? Will their AI system train on our data?

Pillar 3: Vendor Reliability

- Does this AI company possess the **long-term stability** required to support our organization?
- Does the vendor **understand our industry and business size**?
- What are the possible outcomes if an AI vendor is **acquired or shuts down**?

Industry Lesson: *Family Clothing Store*

*Elena** runs a two-location clothing boutique competing against online retailers and chain stores. Before evaluating any AI vendors, she **documented her success criteria**.

Results: Elena's predetermined criteria prevented her from choosing the vendor with the most impressive demo. Instead, she selected the vendor that **best matched her actual business needs**, resulting in a **43% increase in repeat customer sales** and, within 8 months, a **67% improvement in inventory turnover**.

Questions Every AI Vendor Must Answer

When vendors say, "seamless integration," ask them to define exactly what that means and **who pays for it**.

Business Integration Questions:

- "Share **three references from businesses my size** that I can call today."
- "What happens when the system **fails during our busiest day** of the year?"
- "Walk me through implementation when **everything goes wrong**, not when everything goes perfectly."

Financial Protection Questions:

- "What is the **total cost, including all integrations, training, and support** for the first year?"
- "What **percentage of your customers cancel within the first year** and why?"
- "What are my options if this **does not work for my business after six months**?"

Technical Reality Questions:

- "How long does it take employees to become **proficient**?
- What **training and support** do you provide?"
- "What **specific data do you need from our existing systems**, and who migrates that data?"
- "Show me exactly how this **integrates** with [your specific software systems] in a live environment."

The vendors who provide detailed, honest answers to these questions are worth your investment. Those who deflect or offer vague responses are selling promises they cannot keep.

Contract Negotiation Essentials

Many business owners sign vendor contracts without thoroughly reading them. Here are the five terms that matter most:

1. **Performance Guarantees:** Demand specific, measurable commitments with **penalties for non-performance**. Vague promises like "improved efficiency" are worthless.

2. **Integration Support**: Require **vendor responsibility for all system integrations at no additional cost**. If they promise integration, they should guarantee it works.

3. **Data Ownership and Portability:** Ensure you **own your data and can export it easily** if you cancel. Vendors should provide data in standard formats.

4. **Cancellation Terms**: **Negotiate reasonable exit terms with 30-60 days' notice** maximum. Avoid long-term contracts that lock in pricing or features.

5. **Support and Training Commitments:** Specify **included training hours, response times for support, and escalation procedures for problems**.

The Final Decision Framework

After evaluating all vendors, use this checklist before signing any contract:

Business Impact Assessment:
☐ This solution **enhances our customer relationships.**
☐ My team can realistically **use this program effectively within 90 days**.
☐ This **solves actual problems** without creating new issues.

Risk Management Assessment:
☐ We can afford this investment if **ROI takes longer than projected**.
☐ We are protected if the **vendor has issues or gets acquired**.
☐ We can **exit cleanly if the solution does not work** for us.

Strategic Alignment Assessment:
☐ This **advances our long-term business goals**.
☐ This **strengthens our competitive market position**.
☐ This vendor can **grow with our business**.

Check that all nine statements are in alignment with your **AI and technology goals before signing any contract**.

Vendor Selection Reality

Choosing the right vendor takes time, but it is the **key to making AI a game-changer** for your business, not just a costly tool that gathers dust. To succeed with AI, do not just choose AI for its **flashy features**; instead, pick vendors based on **how well they align with your company's overall strategy**.

Paul's Catering almost chose a vendor with a stunning presentation, but reference calls revealed that the system frequently crashed during high-volume periods. The catering company selected a 'less impressive' vendor whose system proved reliable during their busiest wedding season, resulting in **78% improvement in order accuracy** and **zero system failures** during peak periods.

Your AI vendor will become a business partner for years, not just a software supplier. Choose your AI software vendor as if the future of your business depends on it, because it does.

The systematic vendor selection approach **protects your investment** and ensures a successful AI implementation. The next section will show you the TRACK Framework that transforms your chosen technology into a sustainable competitive advantage.

Part II: TRACK

Chapter 6

TRACK Framework

The $22,400 AI Implementation Failure

For more than two years, *David's* manufacturing company tried several times to integrate AI. Each effort **ended in frustration and wasted investment**. First came predictive maintenance software that his team could not understand. Next was customer relationship management AI that **created more confusion than clarity**. His third attempt at inventory optimization AI conflicted with existing processes and was abandoned after six weeks.

David sat in his truck after the third vendor left, staring at the invoice. Twenty-two thousand dollars. His daughter's first year of college tuition, gone. Not on systems that failed, but on systems that were technically flawless, but unused.

"I'm not stupid," he stated with obvious frustration to his wife that night. "I run a successful business. Why can't I make this work?"

That was when *David* realized he had been approaching AI like installing new equipment. But AI is not machinery. It was a change. And change requires a different blueprint entirely.

After discovering the TRACK Framework, the **relief was immediate**. For the first time in two years, David had structure rather than chaos. Each letter represented a checkpoint he could measure, not another expensive guess.

5-Step TRACK Implementation System

Most AI implementations fail because companies **focus solely on technology while missing the human side** of the project's outcome. They select powerful tools without considering employee resistance. They automate processes without informing customers. They chase efficiency gains while accidentally damaging the connections that made their success possible.

The Three Pillar Implementation Protocol from the previous chapter addressed business fit, technical integration, and vendor reliability. **TRACK** builds on that groundwork by simultaneously developing your technology systems, team capabilities, and customer engagement. This approach brings employees, customers, and business partners into your AI advancement as allies rather than obstacles.

The TRACK Framework Plan

Trust (T) establishes the foundation by **building confidence from your employees, customers, and business partners** before implementing any systems. You will create transparent communication plans that explain your AI goals while demonstrating your commitment to maintaining quality personal service.

Ready (R) prepares your **three pillars** (technology systems, employee teams, and customer connections) for coordinated action. You will create guidelines that prioritize service enhancement over pure automation, including training programs that position AI as career advancement opportunities.

Advance (A) launches **synchronized implementation** across all three pillars using 90-day phases. You will implement AI tools in quarterly cycles, conducting feedback surveys between each phase and monthly skill-building sessions for your team.

Connect (C) integrates all three pillars to **strengthen connections**. You will use AI insights to personalize customer interactions while increasing efficiency, leverage employee AI expertise to create mentoring opportunities, and combine customer feedback with employee insights.

Keep (K) maintains momentum through **ongoing optimization and continuous engagement**. You will establish measurement systems that assess both efficiency gains and service quality across all three pillars.

This systematic approach works because AI outcomes depend primarily on how people adopt and optimize the systems, not solely on technical sophistication.

David's TRACK Success

The issue was not the quality of the AI tools *David* selected. The issue was that he lacked a systematic approach. When he discovered the **TRACK** Framework, he realized his failures were caused by having no coordinated strategy.

Using the framework, *David* successfully implemented ***Fishbowl Manufacturing*** with ***Salesforce Manufacturing Cloud*** through a coordinated approach engaging his team and customers.

Results in 12 months: a **67% improvement in production** efficiency, with an **89% reduction in inventory waste**, an increase of 156% in **on-time delivery performance,** and a **94% employee satisfaction** rating for AI-enhanced work processes.

Although *David* faced early setbacks, he eventually achieved success not because of the AI tools but by implementing a better strategy. The critical difference between past failures and later achievements was adopting a thoroughly planned and effectively executed approach.

Although *David* faced early difficulties earlier, he eventually succeeded—not because of AI tools, but thanks to putting a better strategy into action.

Trust: Doing the Groundwork

You cannot fake **Trust**, and it cannot be easily rebuilt once damaged by poor AI implementation. Vendors will pressure you to skip this phase because trust-building does not show up on their project timeline, and they cannot start billing until you pick their software. We have watched this process play out dozens of times. Do not let vendors rush you.

Every successful AI implementation begins with methodical **confidence-building** across employees, customers, and business partners.

Essential **Trust** Building Actions:

☐ Complete employee confidence assessment and address specific concerns directly.

☐ Conduct customer feedback collection and communicate upcoming service enhancements.

☐ Evaluate business partner connections and position AI adoption as local business leadership.

☐ Develop communication plans that maintain quality personal service.

Verification Metrics

When **80 percent of your employees demonstrate confidence** in the AI implementation, and **85 percent of your customers endorse the service enhancements**, plus all significant concerns have been resolved prior to selecting AI systems, you are ready to progress to the next step.

Ready: Preparing Your Three Pillars

Most businesses **skip preparation and jump directly to software selection**, which contributes to implementation failure. **Preparation prevents the chaos** that destroys most AI projects.

Prepare your three pillars for coordinated action without rushing into AI tool selection.

Pillar Preparation Essentials

1. **Technology Pillar:** Establish evaluation criteria **that prioritize service improvement** rather than focusing solely on increased automation.
2. **Team Pillar:** Design training programs that position AI as **career advancement opportunities**.
3. **Customer Pillar:** Create **communication templates and customer feedback collection systems**.

Advance: Putting AI to Work

Launch coordinated implementation across all three pillars using **90-day cycles**. The quarterly cycle prevents them from becoming overwhelmed while providing enough time to see measurable results and make necessary adjustments.

90-Day Implementation Cycle:

Days 1-30:

Build the groundwork with intensive employee training and customer communication.

Days 31-60:

Integrate your systems with current workflows and collaborative team development.

Days 61-90:

Fine-tune systems based on feedback and plan the next cycle.

Connect: Making AI and Personal Service Work Together

Integrate all three pillars to strengthen your personal connections. Utilize AI insights to tailor customer interactions, enhance efficiency, and leverage employee AI expertise while you foster mentoring opportunities.

Exceptional results occur when **systems, team expertise, and customer relationships combine** to deliver service levels beyond what automation or manual work alone can provide.

Keep: Maintaining the Momentum

Maintain momentum through ongoing refinement and continuous engagement. Most businesses celebrate initial AI success and then stop optimizing. That is a mistake. To dominate your market, treat AI implementation as an **ongoing competitive advantage development process**, not a one-time project.

Ongoing Optimization Tracking

1. Conduct **monthly efficiency metrics tracking** progress across all AI implementations.
2. Conduct **quarterly service quality assessments** with employees and customers.
3. **Continuous system refinement** based on usage patterns and feedback.
4. Engage in regular **training updates** to maintain AI proficiency across your team.

Implementation Timeline

- **Months 1-2:** Trust and Ready Phases — **establish strong footing** with defined success criteria and communication plans.
- **Months 3-4:** Advance Phase – **first 90-Day Cycle with regular monitoring and refinement.**
- **Months 5-6:** Connect and Keep Phases — **integrate all three pillars for ongoing refinement and long-term competitive advantage.**

Results: 30-60 days for basic efficiency improvements, 90 days for service enhancement results, and lasting competitive strength.

What Happens Next Is Up to You

The **TRACK** framework transforms AI adoption from expensive experiments into **competitive advantages**. Your connection-based background is not a limitation to overcome—it is your secret weapon that transforms AI implementation from technological catch-up into sustainable market dominance.

You must understand that while this approach is proven and methodical, it is not automatic. It requires **commitment to the process**, especially when vendors try to convince you to skip steps, employees resist changes, or customers ask questions you are not prepared to answer.

The businesses that succeed with **TRACK** are those that commit to doing it correctly over doing it quickly. The businesses that fail are those that try to shortcut the human elements, focusing solely on systems. They want the competitive advantages that AI provides, but they are not willing to do the connection work that makes AI implementation successful.

Your competitors are not waiting for you to finish reading this book. They are implementing AI right now—some correctly, most incorrectly. The question is not whether AI will reshape your industry. The question is whether you will lead that reshaping or scramble to catch up.

The systems exist. The approach is proven. The competitive advantage is available. What happens next is entirely up to you.

The next five chapters cover each **TRACK** phase in detail, beginning with Trust as your baseline. Industry-specific chapters will show how **TRACK** applies within each specialist market.

Chapter 7

Trust(T)

Trust is not built through technology–it is built through transparency.

We have witnessed businesses waste time, money, and talent on AI implementations that failed. The technology was not bad, but the business never addressed the **concerns people**

had, which ultimately determined whether a project **succeeded or failed**.

The **Trust** phase of the framework establishes a baseline for successful AI adoption by **building confidence among your employees, customers, and community** before implementing any systems. **You cannot fake trust**, and it **cannot be easily rebuilt if damaged by poor AI implementation.**

The Trust Crisis That Destroys AI Implementations

*Johnson's Auto Repair** operated successfully for 12 years, serving over 300 regular customers in its suburban location. When owner Mike Johnson decided to implement AI scheduling and customer communication systems, he **focused entirely on the technology, ignoring the importance of trust building**.

Mike installed ***ServiceTitan*** and ***Podium*** **without preparing his team or communicating with customers**. He announced the changes via a brief email: "We're upgrading our systems to serve you better."

The results were catastrophic:

- Within six weeks, **23% of Mike's regular customers stopped returning to the business**.
- Employees **actively sabotaged the new AI systems** by avoiding their use.
- Customer complaints about impersonal service **increased by 178%**.

- In the first quarter after implementation, **revenue dropped by 31%**.

Mike's failure was not technological; both AI systems functioned exactly as designed. **His failure was human**. He implemented AI changes **without first building trust**.

After 18 months of declining business, *Mike* closed *Johnson's Auto Repair*. The technology that was supposed to save his business destroyed it because **never addressed the trust foundation**.

Mike's hardest conversation was not with customers. It was with Ray, his senior mechanic, for fifteen years. "Are you bringing in robots to replace us?" Ray asked, not meeting his eyes.

Mike wanted to lie. Say everything would stay the same. Instead, he sat on the shop floor next to Ray, grease under his fingernails, and said something he had never admitted before. "I don't know what I'm doing, Ray. But I know if we don't figure this out together, we're both out of work in five years."

That honesty changed everything.

AI Implementation Challenge

Every business—regardless of industry—faces the **same three core trust challenges** when implementing AI:

Employees fear job replacement. They worry that automation will eliminate the skilled positions they spent years developing. This fear is universal—from machinists to accountants, from restaurant servers to real estate agents.

Customers worry about losing personal service. They chose your business because of the quality attention that comes from experienced people who understand their specific needs. They fear becoming a number in a system.

The community questions your intent. Business partners, referral sources, and local networks wonder whether AI adoption means abandoning the traditional values and relationships that built your reputation.

These three challenges appear in every AI implementation, regardless of whether you run a machine shop, a restaurant, a law firm, or a farm. **The industry chapters in Part III** of this book will show you how specific businesses addressed these universal concerns within their particular contexts—and what unique concerns each industry adds beyond these three universal concerns.

Industry-Specific Trust Concerns

Different industries face unique **Trust** challenges that require tailored approaches. Supporting implementations with businesses across many different sectors, we have identified specific concerns and strategies that emerge in twelve categories.

Professional Services (Legal, Accounting, Consulting)

Primary Trust Concerns:

- Client confidentiality and data security with AI systems.
- Professional liability and compliance requirements.
- Maintaining the perception of expertise and judgment.
- Concerns about AI replacing professional analysis and insight.

Scripts for Professional Services

Example: Jim runs an eight-attorney legal firm specializing in business law. Before implementing ***Clio with AI*** document review, Jim addressed profession-specific concerns:

Confidentiality Communication: "Our AI tools operate within secure systems and are protected by the same attorney-client privilege that safeguards all our communications."

"AI assists us in reviewing documents more thoroughly, mitigating concerns about confidentiality. It can identify potential issues that might take hours to uncover manually, while our attorneys make strategic decisions based on their legal expertise and your specific situation."

Professional Judgment Emphasis: "AI identifies potential issues that might take hours to find manually, but our attorneys make all strategic decisions based on legal expertise and your specific situation."

Compliance Assurance: All AI implementations have been reviewed by our malpractice carrier and comply with state bar technology standards.

Results: **94% client approval of AI enhancements,** no compliance issues, and a **43% improvement in accuracy of document reviews.**

Healthcare and Wellness Services

Primary Trust Concerns:

- **Patient privacy and HIPAA compliance.**
- **Fear of impersonal medical care.**
- **Concerns about AI diagnostic accuracy.**
- **Anxiety about reduced practitioner attention.**

Scripts for Healthcare and Wellness

Example: *Dr. Martin** runs an integrative health practice with six practitioners. Before implementing ***SimplePractice with AI*** scheduling and client management, she addressed healthcare-specific concerns with her team and patients.

Privacy Protection: "AI tools will help us manage appointments and treatment plans more efficiently while maintaining the same HIPAA protections that safeguard all your health information."

Care Quality Enhancement: "Technology will allow us to spend more time focused on your health goals by handling administrative tasks that previously took time away from your patient care."

Practitioner/Patient Relationship Preservation: "Your relationship with your practitioner remains personal and individualized. AI helps us prepare for your visits and follow up on your progress more effectively."

Results: **91% patient satisfaction** with enhanced scheduling and communication, **zero HIPAA violations**, and a **38% increase in practitioner availability for patient care.**

Restaurants and Hospitality

Primary Trust Concerns:

- Loss of personal hospitality and a welcoming atmosphere.
- Concern AI will interfere with cultural authenticity.
- Fear of staff displacement.

Scripts for Restaurants and Hospitality

Example: *Oath Pizza* started in a 600-square-foot seaside shop on Nantucket Island in 2015. As the craft pizza chain expanded to multiple locations, co-founder *Max Seel* faced a challenge every growing restaurant encounters: how do you implement inventory management technology without **losing the personal attention that made customers fall in love** with your food in the first place?

Seel's approach centered on positioning ***Toast POS*** and its inventory tools as systems that **free staff to focus on hospitality rather than spreadsheets**. "Toast is the hub of all our information," he explains. "Having live access to all that information is what makes Toast so valuable for us." The results—**reducing inventory variances from five percent to below two percent**—came not from forcing technology on resistant staff, but from showing them how it made their jobs easier.

Scripts for Trust-Building at *Oath Pizza*:

For the family member worried about losing the personal touch: "Our AI tools handle inventory tracking and food cost calculations behind the scenes so we can spend more time in the dining room greeting customers and making sure every plate meets our standards. **The technology does the math. We do the hospitality."**

For the longtime employee concerned about learning new systems: "You already know our recipes better than anyone. This system just helps us make sure we never run out of ingredients for the dishes our regulars love. You will still plate every dish the same way you always have. The AI just makes sure the ingredients are there when you need them."

For the customer who values your traditional approach: "We have been serving these same recipes for years, and that is not changing. What is changing is that we now catch when we are running low on ingredients before it becomes a problem. You will never hear us say we are out of your favorite dish because we forgot to order something."

For the staff member worried about being watched or measured: "This is not about tracking your every move. It is about seeing patterns—like noticing we always run short on weekends or that certain ingredients spoil before we use them. When we waste less food, we have more money for things that matter, like better equipment and fair wages."

Results: *Oath Pizza* **reduced inventory variances from five percent to below two percent** by closely adhering to recipes and controlling waste. Managers can now look at a tablet and see exactly how inventory is performing—what food costs look like, whether portions are running over or under—without pulling staff away from customers.

Manufacturing and Trades

Primary Trust Concerns:

- Job security and fears of workforce reduction.
- Safety and quality control issues.
- The decline of craftsmanship and the devaluation of skills.
- Problems with union and worker relations.

Trust Recovery Strategies

Even with careful planning, trust-building efforts sometimes encounter unexpected resistance or fail to address everyone's concerns. **Recovery strategies can salvage AI implementations and strengthen relationships.**

When Employee Trust Fails

Common Failure Patterns:

- **Initial enthusiasm turns to resistance** after employees see actual AI capabilities.
- **Fear increases when employees realize AI can perform tasks** they thought were safe from automation.
- Trust is diminished when the initial deployment of AI results in **increased workload instead of immediate improvement in efficiency.**

Recovery Strategy:

Example: Accounting Firm Recovery. Lisa faced employee resistance after implementing ***QuickBooks AI*** when her 6-person team discovered AI could categorize transactions they thought required human judgment.

> **Step 1: Acknowledge the Problem Directly**. Lisa called a team meeting and said, "I realize the AI capabilities are more extensive than we initially discussed, and I understand why that feels threatening. Let us talk about what this really means for your roles here."
>
> **Step 2: Reframe AI Role with Specificity** "AI categorizes routine transactions, but you review all categorizations and make decisions about complex or unusual items. Your expertise is what ensures accuracy and catches problems AI cannot understand."

Step 3: Create Employee Success Stories
Lisa worked with each team member to identify specific ways AI made their work more interesting and valuable, then shared these stories with the entire team.

Step 4: Adjust Implementation Based on Feedback
Lisa modified AI settings to require human approval for transactions over certain amounts and complex categorizations, giving employees more control over AI decisions.

Recovery Results: Employee satisfaction with AI improved **from 23% to 87% within 8 weeks**, and the team began suggesting optimizations that improved AI performance.

When Customer Trust Fails

Common Failure Patterns:-

- Customers perceive AI implementation as cost-cutting and not as a service improvement.
- Initial AI interactions create frustration or confusion.--
- Communication about AI changes feels impersonal or corporate.

Recovery Strategy Plan:

Example: *Rhodes Insurance Agency,* owner Carl Rhodes* faced customer complaints after implementing ***Applied Epic AI***, as clients felt that initial interactions were less personal than those in previous services.

> **Step 1: Direct Customer Outreach.** Carlos personally called every customer who had expressed concerns to listen to their specific problems and gather detailed feedback.
>
> **Step 2: Adjust AI Implementation Based on Customer Input**. Modified AI settings to route more inquiries to human agents and programmed AI to provide warmer, more conversational responses that matched Carlos's communication style.
>
> **Step 3: Provide Demonstrations Instead of Explanations.** Instead of explaining AI benefits, Carlos showed customers specific improvements: "Your policy research now takes 15 minutes instead of 2 hours, and you can reach us after hours for urgent questions."
>
> **Step 4: Create Customer Success Partnerships.** Invited customers to provide feedback on AI improvements and positioned them as partners in enhancing service quality instead of passive recipients of technology changes.

Recovery Results: Customer satisfaction improved from **67% to 94% within 12 weeks**, and the **complaint volume decreased by 89%** through responsive adjustments.

When Community Partner Trust Fails

Common Failure Patterns:

- Business partners perceive AI adoption as moving away from relationship-based business practices.
- Referral sources worry about changes to service quality or business identity.
- Local business community sees AI implementation as a threat to traditional business values.

Recovery Strategy Principles

Recovery Principle 1: Acknowledge Problems Quickly. Denial or defensiveness makes trust problems worse. Direct acknowledgment and responsibility-taking create a baseline for rebuilding relationships.

Recovery Principle 2: Adjust Implementation Based on Feedback. Willingness to modify AI implementation based on customer and employee concerns demonstrates a genuine commitment to relationship preservation over technology deployment.

Recovery Principle 3: Create Success Partnerships. Transform critics into partners by involving them in solution development and positioning their feedback as a valuable contribution to the improvement process.

Recovery Principle 4: Demonstrate Rather Than Explain. Show the benefits of AI implementation through improved service and results without trying to convince people through explanations of technology capabilities.

Trust recovery requires more effort than initial trust building, but successful recovery often creates stronger relationships than existed before AI implementation challenges occurred.

Employee Trust Assessment and Development

Success Story
Simpson's Hardware Store

*Tom Simpson** runs a family hardware store with eight employees and 40 years of community relationships. Before implementing AI inventory management, Tom conducted systematic employee trust building.

Tom's Trust-Building Process:

Week 1: Individual Conversations – Tom met privately with each employee to discuss AI plans, address specific concerns, and explain how AI would enhance their expertise, not replace their knowledge and skills.

Week 2: Group Training Session – Employees learned about AI benefits and participated in planning the implementation process.

Week 3: Skill Development Planning – Tom created individual development plans showing how AI proficiency would lead to expanded responsibilities and career advancement.

Week 4: Implementation Role Assignment – Each employee received specific responsibilities in the AI implementation process, becoming partners and not victims.

Results after 3 months:

- **97% employee satisfaction with AI implementation.**
- **Zero employee resistance to new systems.**
- **34% improvement in inventory accuracy through enthusiastic adoption.**
- **Employee-driven optimizations improved AI performance beyond vendor expectations.**

Tom's success came from **addressing trust before implementing technology.**

Success Story
Insurance Agency

*Keith** runs an insurance agency serving 185 clients in Washington. Before implementing ***Applied Epic AI***, he conducted intentional customer trust building.

Keith's Communication Strategy:

Phase 1: Direct Outreach – Personal calls to fifty key clients explaining his AI plans and soliciting feedback about potential concerns.

Phase 2: Educational Campaign – Email series explaining how AI would improve business response times while maintaining personal customer relationships.

Phase 3: Transparency Initiative – Open communication about the implementation timeline, expected benefits, and defense of personal service.

Phase 4: Feedback Integration – Monthly surveys allowing customers to provide input on AI implementation effectiveness.

Results after 6 months:

- **92% customer approval** of AI enhancements.
- **48% improvement in customer satisfaction** scores.
- **Zero customer complaints** about reduced personal service.
- **23% increase in referrals** from satisfied clients.

Susan's approach demonstrated that transparent communication about AI implementation builds customer relationships.

Building Trust That Lasts

Trust building creates the baseline that determines whether subsequent phases succeed or fail. Without confidence from your team, customers, and community, even the perfect implementation of technology can cause damage to business relationships and undermine competitive advantages.

Trust-Built Competitive Advantages:

Employee Partnership: Team members become **AI implementation champions** who optimize technology and advocate for continued advancement, instead of resisting or undermining changes.

Customer Loyalty: Clients become **partners in technological evolution**, providing feedback and support, not resisting or fearing service changes.

Community Leadership: The local business network recognizes you as **innovative leaders who strengthen**, instead of posing a threat to community values and relationships.

Trust is not built through promises about AI benefits—it is built through **transparency about AI implementation** that demonstrates a genuine commitment to **preserving and enhancing the relationships that define your business success.**

The next chapter will show exactly how to prepare your implementation across technology, teams, and customer areas to ensure coordinated success that builds on the **Trust** groundwork you have established.

Chapter 8

Ready (R)
Preparing for Success

Proper preparation is essential for successful AI implementation.

After watching some businesses waste hundreds of thousands of dollars on AI implementations that failed due to poor preparation, we can tell you exactly where the problems occur. The **Ready phase transforms trust into planned preparation across technology, team, and customer engagement areas.**

This is not about rushing to implement the first AI tool you find. It is about **designing criteria and processes that ensure coordinated success** across all areas simultaneously.

The Preparation Gap That Destroys AI Implementations

Frank had built trust with his 15-person construction company and 85 clients through open communication about the benefits of AI. His crew was excited about eliminating paperwork, and customers appreciated the commitment to maintaining personal service while improving business efficiency. Everyone supported AI implementation—but *Frank* **skipped the Ready phase and jumped directly to vendor selection.**

The chaos was immediate and devastating. *Frank* chose to implement three different AI solutions from different vendors:

- ***Procore*** for project management coordination.
- ***PlanGrid,*** which organizes and manages blueprints.
- ***Raken*** for daily task reporting and project accountability.

Each tool worked perfectly in isolation, but together they created an operational disaster:

Technology Problems: The three systems were **unable to share data effectively, resulting in duplicate work** rather than eliminating it. Project schedules in ***Procore*** did not sync with blueprints in ***PlanGrid***, and daily reports in ***Raken*** were not connected to either platform.

Team Issues: Employees received **fragmented training separately for each system,** never learning how the systems worked together. Training schedules conflicted with job deadlines, which created stress and resistance to implementation and change.

Customer Complaints: Clients received confusing updates from different systems at varying times, creating the **impression of a chaotic technology experiment** rather than a system improvement.

Results after 3 months:

- **Employee productivity decreased 28%** due to system integration problems.
- **Customer complaints** about inconsistent communication **increased by 72%.**
- Due to integration requirements, **technology costs exceeded the budget by 41%.**
- Frank spent 18 additional hours weekly troubleshooting system conflicts and addressing employee and customer concerns.

The devastating irony: *Frank* had the trust of his employees and customers. **Poor preparation destroyed the implementation before any AI benefits could be realized.**

Frank's project manager, found him at his desk at 10 PM, surrounded by three different laptop screens showing three different systems that refused to talk with each other.

"Remember when construction was about building things?" she asked quietly.

Frank laughed—the bitter kind. "I spent today explaining to a family why their kitchen renovation is delayed because our blueprint software won't sync with our scheduling system. They do not care about our technology problems. They care about Thanksgiving dinner."

That is when *Frank* understood that **preparation is not about the technology working. It is about the business working when technology does not.**

The Three-Area Preparation Plan

Based on tracking successful and failed AI implementations, preparation for the **Ready** phase must occur simultaneously to prevent Frank's failure.

Area 1: Technology Selection Preparation

Develop criteria for assessing and selecting AI tools to function collaboratively.

Essential Technology Preparation:

- Define integration requirements before evaluating any vendors.

- Establish budget parameters, including hidden costs and training expenses.
- Create performance metrics that measure business improvement, not just technology functions.
- Develop rollback procedures for unsuccessful AI implementation.

Success Story
Pet Supplies Plus

The grooming department at *Pet Supplies Plus* faced a problem every pet groomer knows: the phone rings constantly, appointments get double-booked, and the person who should be calming a nervous Schnauzer is instead typing customer information into a computer. When you are elbow-deep in soap and fur, administrative work does not just slow you down—it compromises the care you can give.

The franchise's grooming departments implemented ***Groomer.io***, an AI-driven scheduling platform designed specifically for pet care operations. The system does more than book appointments. It **learns pet preferences**, remembers that Max the Golden Retriever needs extra time because he panics around blow dryers, and **automatically notifies customers** when their pet is due for grooming based on breed-specific timelines.

The AI prevents double-booking by understanding how long different services actually take—not the optimistic estimates owners put on their websites, but real-world timing that accounts for difficult coats, anxious animals, and the inevitable cleanup between appointments. When a customer calls to book a full groom for a Standard Poodle, the system knows that is a three-hour job, not ninety minutes.

What groomers report matters more than efficiency statistics. They describe the AI as **eliminating "administrative drag"**—all those small interruptions that pull attention away from the animal on the table. One groomer put it simply: "I have **more hands-on time with the animals** now." For pets that are anxious or medically sensitive, that focused attention makes a real difference in the quality of care.

The business results followed the improved care. **Scheduling conflicts dropped dramatically. Customer retention improved** because pets received consistent, unhurried attention. And groomers—who got into this business because they love animals, not because they love answering phones—reported higher job satisfaction.

The lesson from Pet Supplies Plus applies to any service business where skilled, hands-on work competes with administrative demands. **AI that handles the paperwork lets your people do what they do best.**

Area 2: Team Preparation

The goal is to design training programs that build employee skills and confidence before AI tools go live, and make clear that **AI makes their jobs better, not smaller.**

Essential Team Preparation:

- **Assess each employee's current comfort level with technology** before designing any training.
- **Create individual learning plans that address specific concerns and skill gaps.**

- **Develop career development pathways** that show how AI proficiency leads to advancement.
- **Establish peer mentoring systems** that pair confident employees with hesitant ones.

Success Story
Rodriguez Marketing Agency

Carlos Rodriguez runs an eight-person marketing agency. Before rolling out ***Hootsuite AI*** for social media management, he spent two weeks on preparing his team, not because his employees were resistant to AI but because he had watched enough implementations fail to know that surprise kills adoption.

Carlos started by having a candid conversation with each team member. He asked two questions: **What worries you about this AI implementation? What would make your job easier?** The answers shaped everything that followed.

His training ran four weeks. The first week focused on the basics: what the AI actually does, what it does not do, and why the agency was adopting it. Carlos was direct: faster scheduling and automated analytics reporting would free the team to focus on strategy and client relationships. Where their expertise actually mattered. He addressed job security concerns directly and on the record. **Nobody was being replaced.**

Weeks two and three were hands-on practice with real client accounts in a test environment. Carlos paired his two most tech-comfortable employees with the three who needed more -time. That **peer mentoring arrangement worked better than any formal training session.**

By week four, the team was customizing the platform and identifying workflow improvements Carlos had not even considered.

Results after six months: Employee satisfaction scores increased 34%. The team's **capacity to manage client accounts grew by 40% without adding staff.** More importantly, three team members who had been hesitant at the start were actively suggesting ways to expand AI use — because they now understood it as a **tool they controlled, not a system replacing them.**

The lesson: your employees will make or break any AI implementation. **Prepare them first, prepare them well, and involve them in the process** rather than presenting them with a finished decision.

Area 3: Customer Engagement Preparation

The goal was to develop communication strategies and feedback systems to position AI as an improvement to personal service.

Essential Customer Preparation:

- Analyze **customer communication preferences** and concerns.
- **Create messaging that emphasizes relationship improvement** over cost reductions.
- **Establish feedback collection systems** for continuous improvement.
- **Design service enhancement demonstrations**, not just technology explanations.

Success Story
Auto Shop

*Carl** runs an auto repair shop serving 300 customers. Before implementing AI diagnostics and appointment scheduling systems, Carl prepared a comprehensive customer engagement strategy.

Carl's Customer Preparation:

Communication Preference Analysis: A survey of 150 customers revealed that **78% preferred phone calls** for important updates about car repairs. **67% wanted text reminders** for appointments, and **89% valued personal explanations** of repair recommendations.

Message Development: Communicated faster diagnoses, accurate estimates, and better availability without compromising personal consultation or quality.

Feedback Integration: Established monthly customer satisfaction surveys and quarterly focus groups to monitor AI implementation impact on customer relationships.

Service Enhancement Demonstrations: Instead of explaining technology features, *Carl* showed customers specific benefits—"Your repair estimate will be ready in 15 minutes instead of 2 hours, and you'll receive text updates about progress."

Results After 8 Months:

- 94% **customer approval of AI enhancements**
- **43% increase in customer referrals**
- **Zero complaints about personal service.**

Carl's preparation helped **AI reinforce, not undermine, customer relationships.**

Preparation Phase Integration and Readiness Verification

Final Readiness Checklist:

Technology Preparation Verification:

☐ All selected AI tools integrate with existing systems without expensive customization.
☐ Total technology costs (including training and support) fit within an established budget.
☐ Performance metrics and success criteria are clearly defined and measurable.
☐ Rollback procedures are documented and tested.

Team Preparation Verification:

☐ All employees have completed the AI readiness assessment (Appendix C) and training preparation.
☐ Individual learning plans address specific employee concerns and skill levels.

☐ Career development pathways incorporate AI proficiency advancement opportunities.
☐ Ongoing support systems are established and ready for implementation launch.

Customer Preparation Verification:

☐ Customer communication preferences and concerns have been thoroughly analyzed.
☐ Messaging strategy emphasizes service enhancement rather than cost reduction.
☐ Feedback collection systems are ready for implementation monitoring.
☐ Key customers have been personally contacted and concerns addressed.

The Ready phase eliminates implementation chaos by creating a plan that encompasses **technology selection, team development, and customer engagement** before any AI systems are deployed.

If you complete thorough preparation, you can **avoid the expensive failures that result from random technology adoption.**

The next chapter will show you exactly how to launch coordinated AI implementation using the 90-day cycle framework that builds on the trust and preparation you have established.

Chapter 9

Advance (A)
Putting AI to Work

The implementation and deployment of AI provides businesses with a **distinct competitive edge**.

The **Advance** phase launches coordinated AI implementation across technology, team, and customers. Engaging in **90-day cycles avoids overwhelming participants while delivering measurable results.**

Many businesses struggle during AI implementation because they lack structured execution methods.

We have found that businesses succeed with AI implementation when they follow **disciplined execution cycles.** Rather than attempting to address all tasks simultaneously or implementing systems without proper coordination.

The Execution Trap That Destroys Implementation

*Baldwins' Accounting & Tax Prep** built trust with its eight employees and 120 clients. They **completed thorough preparation** of the **Advance** phase, preparing their technology, employees, and customers. AI tools were selected and integrated to meet all their criteria. Owner *Terry Baldwin* was confident that AI implementation would proceed smoothly.

Instead of following planned execution cycles, Terry attempted to implement all AI systems simultaneously, including:

- ***QuickBooks AI*** with features that were enabled across all client accounts.
- ***Receipt Bank*** that was deployed for all expense management, and
- ***TaxDome,*** a client portal, was activated for all their customers.

Over the course of an intensive week, employees at Baldwin's underwent AI training. **The program execution was a disaster which unfolded quickly:**

Week 1: The systems worked individually but data integration created conflicts.

Week 2: Employees became overwhelmed trying to learn three new systems simultaneously.

Week 3: Customers received confusing communications from multiple AI-enhanced platforms.

Week 4: Terry spent more than forty hours in one week troubleshooting problems instead of serving clients.

Results after 2 months:

- **Customer satisfaction decreased 34%** due to service disruptions.
- **Employee productivity declined 28%** during an overwhelming transition period.
- Despite excellent preparation, Terry considered abandoning all AI implementation.

Terry sat in her office at 9 PM, adding up the hours she had lost to troubleshooting. Forty hours in one week. Her son's basketball tournament—missed. Her wedding anniversary dinner was postponed. All for systems that were supposed to save time.

"Maybe I'm just not smart enough for this," she told her husband that night.

"You're smart enough to build this business from nothing," he reminded her. "Maybe the problem isn't you. Maybe it's the approach."

The technology that was meant to create a competitive advantage nearly destroyed the business. Terry's failure was not due to poor trust-building or inadequate preparation. Her failure was a **chaotic execution that ignored the systematic implementation principles** successful businesses follow.

The 90-Day Implementation Cycle Framework

Based on tracking successful AI implementations across various business types, organized AI execution requires structured cycles that progressively build AI capabilities. Not attempting to implement everything simultaneously.

Cycle Structure Overview:

- **Days 1-30: Foundation Deployment – Install core AI systems with minimal customer-facing changes** while conducting **intensive employee training** and beginning customer communication.
- **Days 31-60: System Integration – Connect AI tools with existing workflows**, launch advanced team training and development, and **implement customer feedback collection systems.**
- **Days 61-90: Optimization and Assessment – Fine-tune systems based on employee and customer response, measure results against success criteria**, and plan the next cycle.

The Daily Reality: Terry's 90-Day Journey

Day 8: Terry's diary: "Staff meeting disaster. Everyone nods, but nobody understands. I see it in their eyes—the same look my kids had when I tried teaching them calculus. Too much, too fast."

Day 23: "Breakthrough! We figured out how to make Receipt Bank talk to QuickBooks today. She literally danced. It's the first time anyone's smiled in three weeks."

Day 45: "Customer called to thank us for faster service. Did not mention AI. Just said we **seemed 'more on top of things.'** That's when I knew we were getting somewhere."

Day 72: "Team suggested improvements I hadn't considered. **They own this now.** It's not my AI project anymore, **it's our competitive advantage**."

Day 89: "Profit margin up 12%. More importantly, **no one worked past 6 PM** this week. We are not just more efficient. We're more human."

Foundation Deployment Example: *Glori's Landscaping*

*Glori** runs a landscaping company with twelve employees serving 85 residential and commercial clients. Using the 90-day cycle framework, they implemented ***Jobber*** for scheduling and customer management.

Week 1: Technology Baseline

- Installed ***Jobber*** with basic scheduling functionality only.
- **Migrated existing customer data during off-peak hours.**
- Prior to full deployment, tested system integration with existing processes.

Week 2: Team Groundwork

- Conducted **hands-on training with three employees using actual customer scenarios.**
- Established **peer mentoring partnerships** for ongoing support.
- Created reference guides customized to Glori's specific workflows.

Week 3: Customer Communication

- Sent **personalized letters to key clients** explaining upcoming service enhancements.
- Made follow-up phone calls to answer customer questions and address concerns.
- Established feedback collection through monthly customer satisfaction surveys.

Week 4: Platform Consolidation

- Addressed and fixed integration problems found during testing.
- Gathered employee feedback on system usability and training effectiveness.
- Analyzed customer response and refined communication messaging.

Foundation Deployment Results: 89% of **employees have confidence in the new system**. A **94% customer approval rating** of the company's communication approach, **without major technical problems during launch.**

The 90-Day Cycle Success Plan

Cycle 1 (Days 1-90): Foundation Establishment – Build basic AI capabilities with planned team and customer integration.

Cycle 2 (Days 91-180): Capability Expansion – Based on feedback, added advanced AI features while optimizing existing systems.

Cycle 3 (Days 181-270): Competitive Enhancement – Implemented sophisticated AI capabilities that create market differentiation.

Beyond Day 270: Continuous Optimization – Ongoing refinement and expansion based on competitive opportunities and business growth.

Common Implementation Challenges and Solutions

Challenge 1: Issues with Integrating Technology

Bringing new technology into existing systems often presents significant obstacles, such as compatibility problems and resistance to change. These challenges can slow down progress and require careful planning and support to overcome.

> ***Solution*: Test all integrations during the foundation phase before expanding AI functionality.**

Challenge 2: Employee Overwhelm

Excessive hours, multiple projects, and approaching deadlines lead to employee stress.

> ***Solution*: Limit new learning to one major system every 30-day period with ongoing support.**

Challenge 3: Customer Confusion

Customers struggle to understand new systems, products, or services, which leads to uncertainty and questions.

> ***Solution*: Communicate changes before implementation and continuously gather feedback.**

Challenge 4: Difficulty Gathering Results

Collecting data during AI deployment is often hindered by technical limitations. These challenges can impact the reliability of findings and slow the progress of implementing effective solutions.

***Solution*: Establish baseline metrics before implementation begins and measure progress monthly.**

Strategic Groundwork for Sustained AI Success

The **Advance** phase creates systematic implementation discipline that delivers **measurable business improvements while preserving and enhancing relationships** with employees and customers.

Implementation-Built Competitive Advantages:

Execution Excellence: **Disciplined 90-day cycles** ensure AI capabilities are deployed strategically, creating reliable results that compound over time.

Relationship Preservation: Continuous monitoring and adjustment across your technology, employees, and customers ensures AI improves relationships.

Operational Mastery: Coordinated implementation builds internal expertise and optimization expertise that creates **ongoing competitive advantages.**

Sustainable Growth: Systematic deployment lays the groundwork for **continued AI advancement without disrupting business operations.**

The next chapter will show you exactly how to integrate AI capabilities with human expertise and customer insights to create competitive advantages that neither technology-only nor human-only approaches can achieve.

Implementation transforms the possibility of AI from an experiment into a competitive advantage. The systematic approach outlined in this chapter ensures that your AI deployment enhances the relationships that have built your business success.

Chapter 10

Connect (C)

Uniting AI and People

AI does not replace human connections; it is used to improve them.

After tracking successful AI implementations across various industries, we have found that the businesses that **dominate their markets are not necessarily those with**

the most advanced technology. They are the businesses that create systems where **AI makes people more valuable, and people make AI more effective.**

The **Connect** phase integrates technological capabilities, employee expertise, and customer insights to create **service levels that neither human-only nor AI-only can achieve.**

The Integration Gap That Limits AI Potential

Roberto had successfully implemented AI at his marketing agency. ***Hootsuite*** with AI content optimization was managing social media postings with an 87% efficiency improvement. His team was comfortable with AI features, and clients appreciated faster response times and more consistent content delivery.

By every implementation metric, the AI deployment was successful. But Roberto was frustrated because **results plateaued after four months.**

Results after 4 months of successful but disconnected AI usage:

- **Social media posting efficiency improved 87%.**
- **Client satisfaction increased 23%.**
- **Employee confidence with AI systems reached 89%.**
- **Revenue growth stagnated at 12%** despite technological improvements.

Roberto realized the problem: his **AI tools, employees, and customers were operating independently**, not working together to create competitive advantages.

The integration breakthrough came when Roberto connected all three areas:

The magic happened during a Tuesday morning meeting. Julie from Creative was showing AI-generated content insights when Tom from Accounts interrupted, "Wait, that pattern matches what three clients told me last week about their customer complaints."

Roberto watched his team connect dots he had never seen. AI data triggering human insights triggering better AI prompts triggering breakthrough solutions.

"This is it," he thought. "This is what a few hundred dollars a month actually buys—not software, but multiplication. My people times ten."

- **AI insights** were identified, which content types generated the most client engagement.
- **Employee expertise** created sophisticated content strategies based on AI data.
- **Customer feedback** guided AI optimization for better business results.

Results after connecting AI, team, and customers:

- **Revenue growth accelerated to 67%** through AI-enhanced strategic consulting.
- **Client retention improved to 98%** through personalized AI-driven recommendations.

Employee job satisfaction was enhanced by engaging in strategic work as opposed to routine tasks.

Roberto's experience demonstrates why the **Connect phase** transforms good AI implementations into **market-leading competitive advantages.**

Making AI and Human Expertise Work Together

Success Story
TBH Complete Auto Care

Benjamin Hart built his auto repair business on a simple philosophy: transparency and honesty with every customer. When he opened *TBH Complete Auto Care* in San Diego in June 2020, he started with a single bay and big ambitions. Within four years, he had outgrown that space and expanded to a multi-bay facility.

Hart credits ***Mitchell 1 ProDemand*** and ***Manager SE*** with helping him scale without losing the personal touch that made customers trust him in the first place.

"To me, it is not just about putting money in my pocket, but truly **developing relationships with my customers and ensuring their safety on the road**," Hart says.

The diagnostic software lets him quickly attend to vehicles, ensuring a seamless flow in and out of his bays. But what Hart values most is how the integrated ***Manager SE*** system **maintains comprehensive records of every transaction**. This means easy retrieval of service history, work performed, and communication details, so when a repeat customer walks in, Hart and his team already know their vehicle's story.

Results:

- **Expanded from single-bay startup to multi-bay facility in under four years.**
- Seamless customer communication through integrated shop management.

The Takeaway: *Hart's* story shows that **AI-enhanced tools work best when they support what already makes a business successful**, knowing your customers and standing behind your work. **The technology handles the complexity; relationship-building remains personal.**

How to Make AI and People Work Together

Step 1: Identify Integration Opportunities.

Look for situations where AI data can inform human decision-making and where human expertise can improve AI performance.

Step 2: Create Collaboration Workflows.

Design specific processes where AI and human capabilities complement each other.

Step 3: Train Integration Skills.

Develop employee expertise in interpreting AI insights and using AI data to enhance customer relationships.

Step 4: Measure Collaborative Results.

Track outcomes that result from AI-human collaboration, not just technology performance or human activity alone.

Success Measurement Plan

Integration Success Metrics:

- **Collaborative Outcomes**: Results achieved through AI-human teamwork that neither could accomplish independently.
- **Customer Value Enhancement**: Improvements in customer satisfaction attributable to AI-enhanced human expertise.

- **Employee Capability Expansion**: Career advancement and job satisfaction resulting from AI proficiency.

- **How You Standout**: Market advantages created through integrated AI-human capabilities.

Common AI Issues and How to Fix Them

Problem 1: AI and Employees Operating Separately

Solution: Create specific workflows that require AI insights and human interpretation to work together.

Problem 2: Customers Do Not Understand AI Benefits

Solution: **Demonstrate results, not just explaining the technology**, but show faster service, better recommendations, with more personalized attention.

Problem 3: Employees Feel Threatened by AI Capabilities

Solution: Present AI as a tool that **emphasizes employee expertise**, showcasing how it can enhance their value in the workplace instead of replacing them.

Problem 4: AI Implementation Does Not Improve Customer Relationships

Solution: **Use AI insights to personalize customer interactions** and enhance service quality instead of merely boosting efficiency.

The Integration Advantage

Businesses that effectively combine AI capabilities with human expertise and strong customer relationships gain competitive advantages that grow over time. Competitors relying solely on technology cannot replicate the genuine relationships that are enriched by AI.

Conversely, companies with only human resources cannot achieve the same level of efficiency and accuracy in service delivery that AI augmentation additions can provide.

The **Connect phase turns effective AI implementations into lasting market leadership** by integrating technology, people, and customers to work collaboratively.

The next chapter will show you exactly how to maintain and expand these competitive advantages through ongoing optimization and continuous innovation that keeps you ahead of competitors who are just beginning their AI journey.

Chapter 11

Keep (K)

Maintaining Momentum

Most businesses celebrate initial AI success and then stop optimizing. The businesses that dominate their markets are those that **treat AI implementation as an ongoing process for developing a competitive advantage.**

Businesses that fail at achieving long-term AI success are those that **view implementation as a project with a**

defined end date. They achieve initial improvements, declare victory, throw a little party, and then **watch their competitors catch up to them within 18 months.**

The **Keep phase maintains momentum through ongoing optimization and continuous engagement** with your team, customers, and community. This phase determines **whether your AI investment creates temporary improvements or sustainable competitive dominance.**

The Success Trap That Destroys AI Advantages

*Thompson's Landscaping** achieved impressive results during their first year of AI implementation: a **43% improvement in scheduling efficiency, a 67% reduction in customer complaints, and 89% employee satisfaction** with ***Jobber*** scheduling software.

Owner *Mike Thompson* celebrated the success and shifted focus to other business priorities. He **stopped conducting monthly AI optimization reviews, discontinued employee training updates, and reduced the frequency of customer feedback collection.**

What happened next demonstrates why the **Keep** phase is critical:

Mike Thompson remembered the exact moment he stopped trying. It was a Thursday. The AI system was humming along. Customers were happy. Employees had stopped complaining.

"We made it," he told his wife over dinner. "The hard part is over."

Eighteen months later, watching a competitor's truck service the Johnson estate, his best client for eight years, he realized the hard part had just begun. **Success is not a destination. It is a speed you have to maintain.**

Mike's experience shows why the **Keep** phase determines whether AI creates temporary improvements or sustainable competitive dominance.

Month 12-18: Gradual Decline

- Employee enthusiasm declined as they focused on workarounds instead of AI system optimization.

- Exceptional **customer satisfaction scores dropped from 94% to 78%** as service became routine.

- **Competitor implementations caught up, eliminating *Thompson's*** technological advantage.

Month 18-24: Competitive Disadvantage

- Two competitors implemented superior AI systems while *Thompson* stagnated.
- **Employee turnover increased 45%** as team members sought AI-enhanced career opportunities elsewhere.

- **Revenue growth slowed from 34% to 8%** as competitive advantages disappeared.

Month 24: Emergency Intervention

Mike realized that stopping **Keep phase activities had cost him market leadership**. Competitors who started AI implementation 12 months after *Thompson* had surpassed his capabilities through continuous optimization.

The Four Pillars of Sustained AI Success

Pillar 1: Continuous Technology Optimization

Monthly Technology Reviews: Evaluate AI system performance, identify optimization opportunities, and implement improvements based on usage patterns and business results.

Quarterly Capability Expansion: Introduce new AI features and tools that enhance existing capabilities without introducing system complexity.

Annual Strategic Assessment: Review entire AI technology programs for integration improvements, cost optimization, and competitive advancement opportunities.

Pillar 2: Ongoing Team Development

Skills Evolution: Provide regular training on new AI capabilities and advanced optimization techniques that increase employee value and job satisfaction.

Career Advancement: Create promotion pathways based on AI proficiency and innovation that retain talent while building internal expertise.

Innovation Culture: Encourage employees to suggest AI improvements and recognize contributions that enhance business performance.

Pillar 3: Customer Relationship Enhancement

Feedback Integration: Use customer input to guide AI optimization priorities and service improvement initiatives.

Service Evolution: Leverage AI efficiency gains to expand service offerings and improve customer experience.

Loyalty Building: Position ongoing AI improvements as a commitment to customer success.

Pillar 4: Market Leadership Position

Industry Leadership: Share AI implementation success with the business community to establish market position and attract referral opportunities.

Competitive Intelligence: Monitor competitor AI adoption and maintain technology advantages through continuous advancement.

Innovation Communication: Provide frequent updates on service enhancements and technology advancements to customers and the community.

Common Keep Phase Challenges and Solutions

Challenge 1: Technology Complacency

Problem: Systems work well enough, so optimization stops. *Solution*: **Establish monthly review cycles with specific improvement targets.**

Challenge 2: Employee Skill Stagnation

Problem: The Team becomes comfortable with current AI capabilities and stops learning. *Solution*: **Create advancement incentives tied to AI proficiency** and skills development.

Challenge 3: Customer Experience Plateau

Problem: AI improvements no longer translate into better customer experience. *Solution*: **Leverage improvements in efficiency to broaden customer service offerings.**

Challenge 4: Competitive Advantage Erosion

Problem: Competitors catch up while your AI capabilities stagnate. *Solution*: **Implement systems for gathering competitive intelligence** and continually developing capabilities.

The Keep Advantage

The **Keep** phase creates sustainable competitive advantages that strengthen over time:

Technology Leadership: Continuous optimization maintains innovative capabilities that competitors struggle to match.

Innovation Culture: Employee-driven improvement creates advantages that competitors cannot purchase or replicate.

Customer Loyalty: The continuous development of customer relationships helps maintain connections even in the face of competition.

Market Position: Effective community leadership and strong industry influence establish a strategic position that enhances ongoing competitive advantage.

Businesses that successfully implement the **Keep** phase do not just maintain AI advantages—they create competitive positions that become increasingly difficult for competitors to challenge.

Long-Term Success Plan

Year 1: Baseline and Optimization – Develop strong AI capabilities and enhance early-stage project deployments to achieve optimal business outcomes.

Year 2: Expansion and Leadership – Add advanced AI capabilities while establishing market leadership through innovation and results.

Year 3+: Dominance and Influence – Create market positioning that competitors cannot easily challenge through technology alone.

The **TRACK** Framework is complete: **Trust** builds the foundation, the **Ready** phase prepares, **Advance** puts AI to work, **Connect** brings together people and AI, and **Keep ensures your competitive advantages grow stronger every year** by building on the momentum of success.

Sustainable competitive advantage is not about reaching a finish line; it is about creating systems that ensure you keep moving forward faster than anyone can catch up. The **Keep phase transforms temporary AI success into permanent market leadership** that defines your business future.

Chapter 12

T • Trust
R • Ready
A • Advance
C • Connect
K • Keep

Metrics

Measuring What Matters in AI Implementation

Whether you call metrics "Key Performance Indicators" (KPI's), "your numbers," or simply "measurements," you need to **track your progress. Effective management of a**

business requires the ability to measure its evolution or identify areas where growth is lacking.

Failing to monitor progress on an AI project can introduce unnecessary risk to both your financial investment and business operations.

Peter Drucker, the legendary business consultant known as the *Father of Modern Management*, captured a fundamental truth when stating, **"What gets measured, gets managed."**

We know that if you are not monitoring what matters, you are losing control over it.

Picture yourself driving at night with an unlit dashboard. You cannot see the speedometer, fuel gauge, or any dashboard indicators. You are careening forward blind, unable to detect engine problems until smoke pours from the hood or you stop only when you run out of gas. **Companies that fail to track metrics are functioning in a comparable manner.** Racing toward their goals while driving blind to the critical signals that determine the outcome.

Failing to monitor the vital signs of your operation — **cash flow, customer satisfaction, employee productivity, and market share** —means you are not managing your business; you are merely hoping it survives. In today's competitive AI landscape, ***hope is not a strategy.*** **Measurement distinguishes deliberate success from accidental failure.**

Success Story
Accounting Services

*Sandra** had built her accounting practice over twelve years, serving 95 small business clients. When her partner Mike pushed her to implement AI bookkeeping tools, she resisted. "How do you even know the AI is working? Revenue is revenue."

Three months after implementing ***Receipt Bank*** and ***QuickBooks Advanced*** those same spreadsheets told a different story.

Results:

- **Labor hours per tax return decreased by 34%.**
- **Client satisfaction increased 23%.**
- During tax season, *Sandra* was home in time for dinner four nights a week instead of one.

"The numbers do not lie," she told Mike. "I just had to be brave enough to look."

The Common Conversation

This conversation between family and colleagues was happening in thousands of businesses, kitchens, offices, and coffee shops across the country. Entrepreneurs who had built successful operations through hard work and personal relationships were trying to figure out whether AI helped their businesses or just made vendors rich.

Sandra believed in measuring success. She had built up her accounting practice by helping clients understand their numbers. But AI metrics felt different. How do you measure whether artificial intelligence is making your business better without getting lost in technical complexity that has nothing to do with bona fide business results?

AI projects succeed or fail based on numbers because they focus on measurable business improvements. Your AI system learns by evaluating its performance against specific goals. **Without clear metrics, you are essentially asking your AI to aim at a target it cannot see.** The system requires specific feedback for improvement, and it is essential to have clear evidence that it addresses actual problems, rather than merely demonstrating technical features.

Investing in substantial resources, such as talent, technology, and data collection, is vital for the success of AI projects. **Monitoring key metrics** such as system accuracy, technology uptime, team training costs, customer response times, and other business impact indicators is essential. Without these, there is a risk of spending months fine-tuning a solution that performs well in tests but underperforms in real-world scenarios.

We have seen companies waste thousands on AI that, while technically functional, failed to improve revenue or efficiency. The companies lacked the proper measures to track whether they were truly beneficial.

AI also comes with sneaky risks that can turn against you if you are not vigilant. Your AI may start making biased decisions that harm specific customer groups. It could recommend strategies based on outdated data that no longer apply to your market. It may optimize for the wrong goals, making processes faster while accidentally leaving customers less satisfied. These problems can lurk invisible for months before they explode into customer complaints, lost revenue, or regulatory headaches.

The Problem with Traditional Business Metrics

Traditional metrics show past outcomes; AI requires tracking ongoing changes and new possibilities. The standard dashboard of revenue, profit margins, and customer counts might miss the most important AI impacts.

The real measures include **speed of decision-making, reduction in repetitive work, and even employee AI adoption rates.** These are not vanity numbers; they show whether AI is actually creating leverage in your business or just adding another layer of complexity.

Standard measures show financial results too late to change them, but tracking AI metrics allows you to monitor progress in real time. However, not tracking KPI's at all in your business, regardless of its size, can lead to serious disasters.

The Revenue Trap

Revenue growth can hide critical changes that determine long-term success. *Trio Heating and Air* **revenue increased 89%** after implementing ***ServiceTitan AI***, but the real story of its progress was much more complex.

The revenue growth stemmed from three sources: 1) **47% more service calls** resulting from improved scheduling, 2) a **23% increase in average billing values** due to enhanced diagnostic capabilities, and 3) 19% more referrals generated through improved customer appreciation and recommendations. But AI had also identified potential problems that traditional metrics would never reveal.

The system showed that they were **losing 34% of their emergency calls during the peak summer months** because their manual dispatch system could not handle the volume of call spikes. Their best technicians spent 40% of their time on routine maintenance that AI could optimize, leaving less time for the complex repairs that displayed specialists' skills and justified premium pricing.

The revenue showed up four months later when satisfied customers started recommending their company and requesting additional services.

The Efficiency Illusion

Efficiency gains without context can be misleading or even counterproductive. When the auto shop implemented AI diagnostic tools, which **reduced vehicle inspection time from 45 minutes to 12 minutes**, there was a **73% increase in efficiency**. This improvement was a notable enhancement on any dashboard.

But the efficiency metric missed a critical detail: the faster diagnostics created more time for customer consultation and explanation, which improved **customer satisfaction by 56% and increased the success of upselling services by 89%.** The real value was not the time savings; it was what the owner did with the extra time.

Meanwhile, AI diagnostics identified **34% more potential issues than manual inspections,** resulting in more effective preventive maintenance recommendations and fewer customer breakdowns. But focusing solely on efficiency would have overlooked this quality improvement.

The impact on sales showed up four months later when satisfied customers started recommending their company and requesting additional services.

The False Security of Profit Margins

Even profit margins can mask significant business changes. Lisa's restaurant saw **profit margins improve 18%** after implementing ***Toast POS*** with AI inventory management. But Lisa's focus on profit margins almost caused her to miss a bigger opportunity.

The AI system had identified that her weekend brunch crowd preferred lighter menu options, but her staff was still preparing the same heavy dinner portions for brunch service. By the time Lisa noticed customer feedback patterns in her AI analytics, she had missed six months of potential revenue from better menu optimization.

The lesson: Traditional metrics focus on what happened, while AI success requires measuring what is changing and what is possible.

The Four Categories of Metrics That Actually Matter

Smart measurement of AI impact requires tracking **four categories of metrics that work together** to tell the complete story of your business transformation.

1. Efficiency Gains Metrics (Time & Cost)

These metrics show you how AI changes the speed and cost of getting work done. Unlike traditional efficiency measures, AI efficiency gains should create **more time for relationship building and strategic work.**

CUSTOMER RESPONSE TIME

WHAT IT MEASURES:

How quickly you respond to customer inquiries, service requests, or support needs.

WHY IT MATTERS:

Studies show that **46% of customers expect companies to respond within 4 hours**, while 60% define "immediate" as 10 minutes or less.

FORMULA:

`Average Response Time` `=` *`Total Response Time for All Inquiries ÷ Number of Inquiries`*

`Improvement Rate` `=` *`(Old Response Time - New Response Time) ÷ Old Response Time`* `× 100`

REAL EXAMPLE: CUSTOMER RESPONSE TIME

Peterson's Plumbing used AI to improve response time from **4.2 hours** to **47 minutes**.

`Improvement Rate = (4.2 hours - 0.78 hours) ÷ 4.2 hours × 100`

`Improvement Rate =` **`81% improvement`**

TASK COMPLETION SPEED

WHAT IT MEASURES:

How long does it take to complete routine business processes?

WHY IT MATTERS:

Faster task completion frees up time for customer relationship building and strategic work.

FORMULAS:

`Task Completion Rate =` *`Number of Tasks Completed ÷ Time Period`*

`Efficiency Gain =` *`(New Tasks per Hour - Old Tasks per Hour) ÷ Old Tasks per Hour`* `× 100 =` **`% of improvement`**

REAL EXAMPLE: TASK COMPLETION SPEED

> **Green Thumb Landscaping** used AI scheduling to reduce weekly scheduling time from **4 hours** to **1 hour**.
>
> `Efficiency Gain = (`**`40`**` jobs per hour - `**`10`**` jobs per hour) ÷ `**`10`**` jobs per hour × 100 = 300% `**`improvement.`**

LABOR COST OPTIMIZATION

WHAT IT MEASURES:

The actual time and cost savings when AI automation reduces the hours needed to complete business tasks.

WHY IT MATTERS:

Direct labor savings translate to immediate bottom-line impact while freeing employees for higher-value work that drives revenue growth.

FORMULAS:

Labor Hour Savings = (Old Hours per Task × Number of Tasks) - (New Hours per Task × Number of Tasks)

Cost Savings = Labor Hour Savings × Average Hourly Cost.

REAL EXAMPLE: LOST COST OPTIMIZATION

Valley Home Services, an HVAC maintenance company, implemented AI-powered quote generation and invoicing software. Previously, their office manager manually created service quotes and invoices by typing up technician notes, looking up pricing, and calculating totals.

The Numbers:

• Old process: **45 minutes** per quote/invoice (0.75 hours)

• New AI process: **10 minutes** per quote/invoice (0.17 hours)

• Monthly volume: **120 quotes and invoices**

• Office manager hourly cost: **$35** (includes salary, benefits, and overhead)

The Calculation:

`Labor Hour Savings = (0.75 × 120) - (0.17 × 120)`

`Labor Hour Savings = 90 hours - 20 hours =` **`70 hours per month`**

`Cost Savings = 70 hours × $35`

`Cost Savings =` **`$2,450 per month`**

Annual savings: $29,400

The Impact:

Those 70 free hours allowed the office manager to focus on customer follow-up calls and scheduling preventive maintenance appointments, which increased customer retention by 23% and generated an additional $45,000 in annual revenue from scheduled maintenance contracts.

2. Relationship Quality Metrics (The Stuff That Really Matters)

These metrics track whether AI is strengthening or weakening the human connections that built your business.

CUSTOMER SATISFACTION

WHAT IT MEASURES:
How satisfied are customers with your AI-enhanced service?

WHY IT MATTERS:

Studies show that **companies measuring ROI for their AI initiatives are 1.7 times more likely to achieve their goals.**

Track customer satisfaction monthly through surveys using:
Post-service surveys (1-5 scale), can be sent through SMS or via email.

Collect online review scores.

Solicit direct customer feedback.
Sample Questions: (1 = Very dissatisfied, 5 = Very satisfied)
1. Overall Satisfaction:
"On a scale from 1 to 5, how satisfied are you with your overall experience with our company?"
2. Service Quality:
"On a scale from 1 to 5, how would you rate the quality of the product or service you received?"
3. Likelihood to Recommend:
"On a scale from 1 to 5, how likely are you to recommend our company to a friend or colleague?"

FORMULA:

```
Customer Satisfaction Score = Sum of All Ratings ÷ Number of Responses

Satisfaction Improvement = New Score - Old Score
```

REAL EXAMPLE: CUSTOMER SATISFACTION

Riverside Auto Repair implemented an AI chatbot to handle appointment scheduling, service updates, and basic technical questions. They tracked customer satisfaction using a standard 1-5 scale survey sent after each service visit.

Six-Month Comparison:

Before AI implementation:

• Total ratings: **1,823 points** from **485 responses**
• Average score: 1,823 ÷ 485 = **3.76** out of 5

After AI implementation:

• Total ratings: **2,387 points** from **523 responses**
• Average score: 2,387 ÷ 523 = **4.56** out of 5

```
Satisfaction Improvement = 4.56 - 3.76

Satisfaction Improvement = +0.80 points (21% increase)
```

The Impact:

The AI chatbot provided instant responses 24/7, reducing waiting times and customer frustration. The improved satisfaction scores correlated with a 34% increase in repeat customers and a 28% boost in positive online reviews, contributing to a $127,000 increase in annual revenue.

EMPLOYEE HAPPINESS INDEX

WHAT IT MEASURES:
Whether AI is making work more or less enjoyable for your team.

WHY IT MATTERS:
Happy employees provide better customer service and stay longer. Studies show that companies with high employee satisfaction see **31% lower turnover** and **20% higher productivity**.

FORMULA:

```
Employee Happiness Index = (Job Satisfaction
Cumulative Score + Work-Life Balance
Cumulative Score + Technology Confidence
Cumulative Score) ÷ 3
```

Calculate monthly using anonymous surveys rating 1-5 scale

Component 1: Job Satisfaction

• **Meaningful work time** - Percentage of day spent on valuable tasks vs. routine busywork

• **Skill utilization** - Whether AI enhances or undermines employee expertise

• **Achievement satisfaction** - Feeling of accomplishment at the end of the workday

• **Growth opportunities** - New skills and responsibilities AI creates

Example Questions:

1. *How meaningful does your daily work feel?*
   (1 = not meaningful – 5 = extremely meaningful)
2. *How often are you able to use your expertise and skills?*
   (1 = very rarely – 5 = very often)
3. *How satisfied are you with your accomplishments each day?*
   (1 = rarely satisfied – 5 = exceptionally satisfied)
4. *How much are you learning and growing in your role?*
   (1 = not very often – 5 = quite often)

Component 2: Work-Life Balance

• **Overtime frequency** - Whether AI reduces or increases extra hours

• **Stress levels** - Job-related anxiety and pressure

• **Evening work** - Time spent on business tasks after hours

• **Weekend availability** - Boundary preservation during personal time

Example Questions:

1. *Since using AI tools, are you working less time than before?*
   (1 =significantly more time – 5 = significantly less time)
2. *How has AI impacted your overall work stress levels?*
   (1 = much higher stress – 5 = much lower stress)
3. *How has AI affected how much work you do after hours?*
   (1 = much more work – 5 = much less work)
4. *Since integrating AI, how much more are you required to work on weekends?*
   (1 = much more often – 5 = much less often)

Component 3: Technology Confidence

- **Confidence using AI systems** - Comfort level with new technology
- **Feeling valued and productive** - Sense that AI enhances rather than replaces their contribution

Example Questions:

1. *What is your comfort level with the AI technology? (1 = not comfortable at all – 5 =extremely comfortable)*
2. AI allows me to complete my work more efficiently than before. (*1 = strongly disagree – 5 = strongly agree)*

Real Example: EMPLOYEE HAPPINESS INDEX

Bright Horizons Dental Practice implemented AI scheduling software and automated patient record management. They conducted monthly anonymous surveys with their 12-person team to track the impact on employee satisfaction.

Employee Example: Sarah Manning, Office Manager

Before AI Implementation (3 months average):

- Job Satisfaction: **2.8** (spent 60% of day on scheduling conflicts and data entry)
- Work-Life Balance: **3.2** (frequent evening calls to resolve scheduling issues)
- Technology Confidence: **3.3** (comfortable with existing systems)

```
Baseline Index: (2.8 + 3.2 + 3.3) ÷ 3 = 3.10
```

After 90 Days of AI Implementation:

- Job Satisfaction: **4.5** (now spends 70% of day on patient relationships and practice improvement)

• Work-Life Balance: **4.4** (AI handles after-hours appointment requests automatically)
• Technology Confidence: **4.3** (gained skills in AI system management)

```
New Index: (4.5 + 4.4 + 4.3) ÷ 3 = 4.40

Total Improvement: +1.30 points (42% increase)
```

The Impact:
Over the next six months, the practice saw zero employee turnover (previously averaging 2 departures per year), a 47% reduction in patient scheduling errors, and received consistently higher patient satisfaction scores due to more attentive staff. The improved employee morale translated into better patient care and $89,000 in reduced recruiting and training costs.

CUSTOMER RETENTION RATE

WHAT IT MEASURES:
Whether AI helps you keep customers longer.

WHY IT MATTERS:
Retaining customers costs **5-25 times less** than acquiring new ones.

FORMULA:

```
Customer Retention Rate = (Customers at End of Period - New Customers) ÷ Customers at Start of Period × 100
```

EXAMPLE: CUSTOMER RETENTION RATE

Started with **100 customers**, gained **fifteen new customers** ended with **108 total customers.**

```
Retention Rate = (108 - 15) ÷ 100 × 100
Retention Rate = 93%
```

Referral Generation Rate

WHAT IT MEASURES:

How often are satisfied customers recommended to your business?

WHY IT MATTERS:

Referrals indicate strong relationships and trust. Referred customers have a **37% higher retention rate** and **16% higher lifetime value**.

FORMULA:

```
Referral Rate = Number of New Customers from
Referrals ÷ Total Existing Customers × 100
```

TRACK REFERRAL GENERATION RATE MONTHLY:

- Ask new customers how they found you
- Track referral sources in your CRM

EXAMPLE: REFERRAL GENERATION RATE

Sunset Veterinary Clinic implemented an AI follow-up system that automatically sent personalized care reminders and pet health tips after appointments.

Monthly Tracking Results:

- Existing customers: **325 active clients**
- New customers from referrals: **42 in the month**

```
Referral Rate = 42 ÷ 325 × 100
Referral Rate = 12.9%
```

The Impact:

Before the AI system, the clinic averaged 5% monthly referral rate. The improved rate of 12.9% represents a 158% increase in word-of-mouth growth. Since referred customers cost nothing to acquire and typically have higher lifetime value, this translates to an additional $67,000 in annual revenue with zero marketing spend.

3. Revenue Impact (Show Me the Money)

These metrics connect AI improvements to actual financial results but measure them in ways that show AI's unique contributions.

Revenue per Customer

WHAT IT MEASURES:
How much revenue each customer generates over time.

WHY IT MATTERS:
AI should help you serve customers better, leading to increased spending.

FORMULA:

```
Revenue per Customer = Total Revenue ÷
Number of Active Customers

Growth Rate = (New Revenue per Customer -
Old Revenue per Customer) ÷ Old Revenue per
Customer × 100
```

REAL EXAMPLE: REVENUE PER CUSTOMER

Manuel's Restaurant implemented Toast AI, increasing average customer spending from **$24** to **$31** through better recommendation systems.

```
Growth Rate = ($31 - $24) ÷ $24 × 100
```

```
Growth Rate = 29% improvement
```

First-Time Customer Conversion Rate

WHAT IT MEASURES:

How often do potential customers become actual customers?

WHY IT MATTERS:

AI should improve your ability to convert inquiries into sales.

FORMULA:

Conversion Rate = *Number of New Customers ÷ Number of Inquiries* × 100

Improvement = *New Conversion Rate - Old Conversion Rate*

EXAMPLE:

Coastal Insurance Agency implemented an AI chatbot to handle initial inquiries and schedule consultations with agents. The system provided instant responses 24/7 and qualified leads before routing them to the appropriate specialist.

Before AI:

- Monthly inquiries: **180**
- New customers: **27**

```
Old Conversion Rate = 27 ÷ 180 × 100 = 15%
```

After AI:

- Monthly inquiries: **210** (more leads engaged due to 24/7 availability)
- New customers: **55**

```
New Conversion Rate = 55 ÷ 210 × 100 = 26%

Improvement = 26% - 15% = +11 percentage points
```

The Impact:

The AI chatbot captured leads outside business hours (38% of inquiries came between 6 PM and 9 AM) that would have been lost otherwise. The improved conversion rate generated an additional 28 customers per month, resulting in $156,000 in additional annual premium revenue.

Project Profit Margin Enhancement

WHAT IT MEASURES:
How AI improves profitability per project or service.

WHY IT MATTERS:
Better margins create resources for growth and relationship investment.

FORMULA:

```
Profit Margin = (Revenue - Costs) ÷ Revenue × 100

Margin Improvement = New Profit Margin - Old Profit Margin
```

EXAMPLE: PROJECT PROFIT MARGIN ENHANCEMENT

Summit Digital Marketing Agency implemented AI tools for content creation, social media scheduling, and campaign optimization. The automation reduced the hours needed per client project while maintaining quality.

Typical Monthly Client Project:

Before AI Implementation:

- Project revenue: **$5,000**
- Labor costs: **$2,800** (70 hours at $40/hour)
- Other costs: **$700** (software, ads, tools)
- Total costs: **$3,500**

```
Old Profit Margin = ($5,000 - $3,500) ÷ $5,000 × 100 = 30%
```

After AI Implementation:

- Project revenue: **$5,000** (same pricing)
- Labor costs: **$1,600** (40 hours at $40/hour)
- Other costs: **$800** (added AI subscription, reduced other tools)
- Total costs: **$2,400**

```
New Profit Margin = ($5,000 - $2,400) ÷ $5,000 × 100 = 52%
```

```
Margin Improvement = 52% - 30% = +22 percentage points
```

The Impact:

With 25 active clients per month, the improved margins added $27,500 in monthly profit ($330,000 annually). The agency reinvested these funds in hiring a client success manager and offering premium services, which attracted higher-value clients and further increased revenue.

4. Operational Excellence (Making Everything Work Better)

These metrics track whether AI is making your business operations more reliable and predictable.

System Uptime and Reliability

WHAT IT MEASURES:
How often do your AI systems work when you need them?

WHY IT MATTERS:
Unreliable AI creates customer service problems and employee frustration.

FORMULA:

```
System Uptime = (Total Time - Downtime) ÷
Total Time × 100
```

Target: 99%+ uptime for customer-facing AI systems

EXAMPLE: SYSTEM UPTIME & RELIABILITY

Harbor Medical Clinic implemented an AI appointment scheduling system integrated with their patient management software. They tracked system availability over a 90-day period to ensure reliability for both staff and patients.

90-Day Tracking Period:

- Total time: **2,160 hours** (90 days × 24 hours)
- Downtime: **8 hours** (scheduled maintenance: 6 hours, unplanned outage: 2 hours)

```
System Uptime = (2,160 - 8) ÷ 2,160 × 100
```

```
System Uptime = 99.63%
```

The Impact:

The 99.63% uptime exceeded the clinic's 99% target, ensuring patients could book appointments reliably around the clock. During the brief 2-hour unplanned outage, the system automatically switched to a backup phone line, preventing appointment loss. The high reliability built patient confidence in the digital system, resulting in 73% of appointments being self-scheduled online rather than requiring staff time.

Error Rate Reduction

WHAT IT MEASURES:

How AI reduces mistakes in routine business processes.

WHY IT MATTERS:

Fewer errors mean better customer experiences and lower fixing costs.

FORMULA:

```
Error Rate = Number of Errors ÷ Total
Transactions × 100

Error Reduction = Old Error Rate - New Error
Rate
```

EXAMPLE: ERROR RATE REDUCTION

Precision Parts Manufacturing implemented AI-powered quality control inspection using computer vision to detect defects in manufactured components. The system replaced manual visual inspection for initial screening.

Monthly Comparison (5,000 parts inspected):

Before AI (Manual Inspection):

• Total parts inspected: **5,000**

• Defective parts missed: **87** (caught by customers or later quality checks)

`Old Error Rate = 87 ÷ 5,000 × 100 =` **`1.74%`**

After AI Implementation:

• Total parts inspected: **5,000**

• Defective parts missed: **23** (human reviewer double-checks flagged parts)

`New Error Rate = 23 ÷ 5,000 × 100 =` **`0.46%`**

`Error Reduction = 1.74% - 0.46% =` **`1.28 percentage points (74% reduction)`**

The Impact:

The 74% error reduction prevented defective parts from reaching customers, saving an average of $850 per error in warranty claims, returns, and customer relationship repair. Monthly savings totaled $54,400. Additionally, customer complaints dropped by 68%, and the company's quality reputation improved significantly, leading to three new major contracts worth $2.1 million annually.

Process Consistency Improvement

WHAT IT MEASURES:

How AI makes business processes more standardized and reliable.

WHY IT MATTERS:

Consistent processes create predictable customer experiences.

TRACK THROUGH:

- Customer complaint rates
- Quality control metrics
- Employee process adherence & Rework frequency

EXAMPLE: PROCESS CONSISTENCY IMPROVEMENT

Lakeside Hotel Group implemented an AI-guided housekeeping checklist system that provides staff with step-by-step room preparation instructions via tablet. The system ensures every room meets the same quality standards regardless of which team member cleans it.

Six-Month Comparison Across 4 Properties:

Before AI Implementation:

Customer complaint rate: 3.8% of guests (mostly about inconsistent room cleanliness)
Rework frequency: 12% of rooms required re-cleaning before guest check-in
Quality control: 73% of rooms passed initial inspection
Process adherence: Wide variation based on staff experience and training

After AI Implementation:

Customer complaint rate: 0.9% of guests (**76% reduction**)
Rework frequency: 3% of rooms (**75% reduction**)
Quality control: 96% of rooms passed initial inspection (**+23 points**)
Process adherence: Standardized - new staff now match veteran performance within 2 weeks

The Impact:

The consistent guest experience led to a 2.3-point increase in online review ratings (from 3.9 to 4.2 stars). This improvement generated an additional 847 bookings over six months, worth $183,000 in room revenue. Training time for new housekeepers decreased from 6 weeks to 2 weeks, saving $12,000 annually in training costs. Most importantly, staff turnover in housekeeping dropped by 41% as employees felt more confident and less stressed about meeting quality standards.

90-Day AI Metrics Implementation

Instead of attempting to monitor every metric, all at one time, implement metrics systematically to build a measurement discipline without overwhelming your team.

Days 1-30: Baseline Establishment

Week 1: Core Business Metrics

1. ***Average Response Time*** **-** Track how long it takes to respond to customer inquiries.
    - Use existing email timestamps or phone logs.
    - Record daily for one week to establish a baseline.
2. ***Customer Satisfaction Score*** **-** Send a 5-question survey to recent customers.
    - Use a 1-5 scale for overall satisfaction.
    - Track weekly average.
3. ***Revenue per Customer*** **-** Calculate total monthly revenue ÷ number of active customers. Compare this month to the same month last year.

Week 2: Add Efficiency Tracking

4. ***Task Completion Time*** **-** Pick your most time-consuming routine task.
    - Track time daily before AI implementation.
    - Use a phone timer or a basic time tracking app.

Week 3: Employee Perspective

5. ***Employee Satisfaction Check*** – Track an anonymous 3-question survey.

    - Rate current workload, stress level, job satisfaction (1-5 scale).
    - Conduct weekly meetings during the transition period.

Week 4: Document Everything - Create a simple spreadsheet with daily/weekly entries.

- Note any major business events that might affect metrics.
- Take screenshots of existing system performance for comparison.

Days 31-60: Implementation Tracking

Week 5-6: AI System Deployment - Continue tracking all baseline metrics.

Add AI system performance metrics:

- System uptime
- AI accuracy rate (if applicable)
- Training time required for the team

Week 7-8: Early Impact Assessment - Calculate changes in core metrics.

- Document unexpected benefits or challenges.
- Adjust tracking methods based on early results.

Days 61-90: Comprehensive Analysis

Week 9-10: Relationship Impact. Add advanced relationship metrics:

- Customer retention rate.
- Referral generation rate.
- Employee happiness index.

Week 11-12: Financial Analysis. Calculate the complete ROI using all tracked metrics:

- Operating margin improvement.
- Cash flow consistency.
- Total cost of AI implementation vs. total benefits.

Simple Metrics Template

Create a monthly tracking sheet with these columns:

Metric	Baseline	Month 1	Month 2	Month 3	Goal	Variance
Avg Response Time (hours)	4.2	3.8	1.2	0.9	<1.0	-78%
Customer Satisfaction (1-5)	3.8	3.9	4.3	4.6	>4.5	+21%
Revenue per Customer	$240	$245	$278	$310	>$300	+29%
Employee Satisfaction (1-5)	3.2	3.3	3.8	4.1	>4.0	+28%
Monthly Profit Margin %	18%	19%	23%	26%	>25%	+44%

Industry-Specific Metrics

Different businesses should emphasize different metrics based on their core success factors and customer expectations.

Service Businesses (HVAC, Plumbing, Electrical)

Primary Metrics:

- First-call resolution rate (AI diagnostics should improve problem-solving accuracy).
- Emergency response time (AI scheduling should optimize urgent calls).
- Customer referral rate (better service should generate more word-of-mouth).
- Technician productivity (AI should free up time for complex repairs).

Secondary Metrics:

- Parts inventory accuracy (AI should reduce stockouts and overstock).
- Service call duration (efficiency gains without rushing customers).
- Safety incident rate (AI monitoring should enhance workplace safety).

Professional Services (Legal, Accounting, Consulting)

Primary Metrics:

- Client retention rate (AI should enable better advisory services).

- Project completion accuracy (AI research should improve outcomes).
- Revenue per client (enhanced capabilities should justify higher fees).
- Employee billable hour productivity (AI should eliminate non-billable research).

Secondary Metrics:

- Client satisfaction with communication (AI should enable more responsive service).
- Proposal win rate (AI insights should improve competitive positioning).
- Compliance accuracy (AI should reduce regulatory errors).

Retail and E-commerce

Primary Metrics:

- Customer lifetime value (AI personalization should increase loyalty).
- Inventory turnover rate (AI should optimize stock levels).
- Conversion rate improvement (AI recommendations should boost sales).
- Customer service response time (AI should enable instant support).

Secondary Metrics:

- Return rate (AI should improve purchase accuracy).
- Cross-selling success (AI should identify upselling opportunities).
- Seasonal demand forecasting accuracy (AI should predict trends).

Manufacturing and Trades

Primary Metrics:

- Product quality consistency (AI should reduce defects).
- Equipment uptime (AI maintenance should prevent failures).
- Safety incident rate (AI monitoring should improve workplace safety).
- Customer delivery time (AI scheduling should improve reliability).

Secondary Metrics:

- Material waste reduction (AI optimization should minimize scrap).
- AI should support, not replace, employee skill development.
- Customer satisfaction with product quality.

Creating Your Personalized AI Metrics Dashboard

Build a measurement system that fits your business size, industry, and growth stage instead of trying to track everything.

Small Business Dashboard (1-10 employees)

Track Monthly:

- Customer response time
- Customer satisfaction score
- Revenue per customer
- Employee happiness index
- One operational metric specific to your industry

Track Quarterly:

- Customer retention rate
- Referral generation rate
- AI system ROI calculation
- Competitive positioning assessment

Medium Business Dashboard (11-50 employees)

Track Weekly:

- Customer response time
- System uptime
- Employee productivity metrics

Track Monthly:

- Customer satisfaction and retention
- Revenue impact metrics
- Process consistency scores
- Staff training progress

Track Quarterly:

- Comprehensive ROI analysis
- Competitive advantage assessment (see addendum)
- Strategic goal alignment review (see addendum)

Growing Business Dashboard (50+ employees)

Track Daily:

- System performance and uptime
- Customer service metrics
- Key operational indicators

Track Weekly:

- Budget impact metrics
- Employee satisfaction scores
- Process improvement opportunities

Track Monthly:

- Strategic business impact analysis
- Market positioning assessment
- Innovation opportunity identification

Common Measurement Mistakes to Avoid

Mistake 1: Measuring Too Many Things at Once

Problem: Overwhelming your team with dozens of metrics creates confusion, not clarity.

Solution: Begin with 3-5 key metrics and include additional ones only after developing consistent improvements.

Mistake 2: Focusing Only on Cost Savings

Problem: Cost reduction metrics miss revenue generation and relationship improvement opportunities.

Solution: Balance efficiency metrics with growth and satisfaction measurements.

Mistake 3: Ignoring Baseline Establishment

Problem: Without proper before/after comparison, you cannot prove AI impact.

Solution: Spend 30 days establishing reliable baseline measurements before AI implementation.

Mistake 4: Measuring Technology Instead of Business Impact

Problem: Tracking AI accuracy rates and processing speeds instead of customer satisfaction and revenue impact.

Solution: Always connect technical metrics to business outcomes that matter to your success.

Mistake 5: Not Sharing Metrics with Your Team

Problem: Keeping measurement results private limits employee engagement and improvement opportunities.

Solution: Share monthly metric summaries with staff and celebrate improvements together.

The Metrics Success Mindset

Successful AI measurement requires treating metrics as strategic tools, not as an administrative burden. To create sustainable competitive advantages through AI, you must build measurement discipline into your culture instead of viewing it as separate from daily operations.

Measurement is not about proving AI success; it is about ensuring AI success through continuous improvement based on genuine business data.

Your AI metrics dashboard becomes your early warning system for problems and your spotlight for opportunities. When customer satisfaction starts declining before revenue drops, you can fix relationship issues before they become financial problems. When employee happiness increases alongside productivity, you know AI is enhancing and not threatening your workplace culture.

Most importantly, metrics provide the evidence you need to make confident decisions about expanding AI capabilities, adjusting implementation approaches, or pivoting to different AI tools based on actual performance.

The goal is not perfect measurement—it is better management through systematic tracking of what matters most to your business success.

Sandra never became a data scientist, but she became a much better business owner by measuring what mattered. Six months after implementing AI bookkeeping tools with proper metrics tracking, she had proof that technology was making her accounting practice more successful: **34% faster client service, 67% less evening work, 89% better client satisfaction, and a 23% increase in revenue per client.**

"I still do not understand how all the AI algorithms work," Sandra told Mike over dinner. "But I understand perfectly how our business is working better because of them. And that is all I need to know to make smart decisions about our future."

Your AI transformation success depends on the same practical approach: measure what matters, track what changes, and use real data to guide decisions that make your business stronger and not just more efficient.

The metrics you track today determine the competitive advantages you build tomorrow.

Part III: Innovate

Chapter 13

Manufacturing

Making Craftsmanship More Powerful with AI

Manufacturing with AI is not about eliminating skilled workers; it is about giving them superpowers. From **machining, cabinet makers,** to **metal fabrication** and beyond, when we talk to manufacturers about AI, they often picture robots taking over their shop floors and computers making decisions that should come from skilled tradespeople with years of experience. That is not what AI implementation looks like. **The manufacturers winning with AI are not replacing their workers; they are turning them into AI-enhanced experts who can do things that seemed impossible just a few years ago.**

Success Story
Marten's Machining

Walking into *Al Marten's Machine Shop* in Stevens Point, Wisconsin, prior to AI integration you would see one of his master machinists set up a complex job. This guy had 20 years of experience and could estimate tolerances that most people require calipers to measure. He said flat out: no computer could understand what he knew about metal, tools, and the way materials behave under stress.

Three months later, that same machinist gave us a tour of jobs he was running with **AI-assisted programming**. He showed us parts with tolerances he would have turned away before—not because he lacked the skill, but because the setup calculations would have taken longer than the job was worth. **The AI handled the math in minutes.** The machinist handled everything that actually mattered: understanding when something was not quite right, making judgment calls about tool selection, and ensuring quality that no computer could verify on its own.

Al Marten's career demonstrates how skilled artisans can effectively integrate technology into their work while preserving their core values and craftsmanship. In 1984, *Al Marten* opened *Marten's Machining*. With nothing more than a band saw, manual mill, and a lathe, he successfully operated out of his two-car garage.

Three years later Marten faced a defining moment. He needed better precision for a job machining stainless steel cases used for printing checks. Manual equipment could not achieve the required tolerances. Instead of losing the contract, Marten invested in his first CNC machine, a leap of faith that cost more than his garage shop was worth.

Today, *Marten's Machining* operates:

- A team of five full-time machinists using **AI-driven computer-controlled programming** to enhance efficiency.

- **Advanced computer-controlled equipment** guided by smart optimization systems.

- **Predictive maintenance AI** that prevents costly equipment failures.

- **Computer-aided design with AI assistance** for complex medical device component manufacture.

- Quality control systems that ensure **precision beyond human measurement capability**.

The crucial AI insight: Marten never stopped being a craftsman. He became a craftsman with AI-enhanced capabilities that let him tackle jobs requiring precision beyond what manual methods could ever achieve.

That is what AI in manufacturing actually looks like. Not robots replacing craftsmen. Not computers making decisions that require years of experience to get right. It is skilled workers with capabilities they did not have before.

The AI Implementation Challenge for Manufacturing

Manufacturing adds unique concerns for AI implementation. **Skilled workers worry about the devaluation of their craftsmanship, safety managers fear that automated systems might miss critical safety considerations, and quality control specialists question whether AI can maintain the precision standards that define their reputation.**

The Manufacturing Transformation Challenge

Manufacturing faces unique AI adoption challenges. **Success depends entirely on bona fide skills and the ability to solve complex problems with materials and site conditions that no computer has experienced.** Expert craftsmen have built careers from years of firsthand experience dealing with real-world challenges that textbooks cannot teach.

Yet, **customers increasingly demand shorter lead times, tighter tolerances, and transparent project tracking, which manual processes struggle to deliver.** This is not about choosing between traditional craftsmanship and modern technology; it is about using AI to make skilled workers more capable, valuable, and competitive than before.

Trust: Building Craftsmen's Confidence in AI Enhancement

A major challenge in implementing AI for manufacturing is addressing concerns among experienced workers about the impact of AI on their professional expertise and its relevance in the workplace.

Trust-Building Success

Owner Mike faced this exact challenge at his precision machine shop. His master machinists, each with 15-20 years of experience, took pride in their ability to set up complex jobs without computer assistance.

Scripts for Personal Conversations That Made the Difference:

For the senior machinist concerned about job security: "This AI programming assistance means you will spend your time on the complex setups that showcase your expertise instead of the routine calculations that any computer can do. You are the one with 20 years of experience who understands when something is not quite right—the AI just handles the math."

For the experienced operator worried about learning new systems: "You already understand machining better than anyone. This is just giving you better tools, like upgrading from a manual calculator to using formulas in a spreadsheet. Your knowledge becomes more powerful, not less important."

For the craftsman concerned about quality: "You still make every quality decision. The AI just helps you identify potential problems before they happen, like having a really smart apprentice who never gets tired and catches details you might miss when you are focused on the big picture."

Results:

- 100% of experienced machinists agreed to try the **AI programming assistance**.
- **Technology decisions** scored an average confidence rating of 4.3 out of 5.
- 89% of machinists **believed AI would make their skills more valuable**.
- Employees had zero resistance to training programs.

Ready: Preparing Systems and People

Vendors will not mention during their slick demonstrations that **manufacturing AI fails when it meets messy reality.** That CNC optimization software looks brilliant when the salesperson runs it on their demo machine with perfect tooling and pristine material. It looks considerably less brilliant when your shop is running three different generations of equipment, your tooling inventory has not been audited in two years, and half your job files live on a computer running Windows 7.

Before you buy any AI system, walk your shop floor with fresh eyes. How many of your machines can actually connect to a network? We have visited shops where owners bought predictive maintenance AI only to discover that their oldest, and most critical, equipment had no way to transmit data. **They ended up with an expensive monitoring system that covered 60% of their production capacity while their vintage Bridgeport sat in the corner, unmeasured and unpredictable as ever.**

Your data situation matters more than your equipment age. AI learns from information, and manufacturing AI needs a lot of data from cycle times, tool wear rates, material specifications, quality inspection results, and maintenance records. Pull out your documentation from the last year. Can you decipher the average setup time for your ten most common jobs? Do you know which materials create the most scrap? If your production data lives in your machinists' heads or on handwritten job tickets that get tossed at the end of each shift, that is your real starting point, not the AI purchase.

Talk to your craftsmen before you talk to vendors. Have one-on-one conversations where you ask what frustrates them about current processes, what calculations eat up their time, and what information they wish they had at the machine. The machinists who have been running your equipment for twenty years know exactly where AI could help, and exactly where AI would get in the way. **Their input should shape your requirements, not the other way around.**

Get honest about your maintenance practices. Predictive maintenance AI promises to tell you when equipment will fail before it happens. That sounds like magic until you realize the system needs baseline data about normal operation to detect abnormalities. If your preventive maintenance schedule is more theoretical than actual, if your maintenance logs are incomplete, if nobody really knows what "normal" sounds like on that old lathe, **the AI will spend months learning what your experienced operators already know by ear.** Sometimes, preparation means finally tackling the basic tasks you have delayed, not buying new technology.

Finally, know your tolerance for disruption. Manufacturing runs on schedules, and AI implementation means learning curves, troubleshooting, and the occasional system that does not work as advertised. **The shops that succeed carve out time for experimentation without betting production deadlines on unproven technology.** Start with a single machine, a single process, or a single product line. Prove the value before you scale. **Your craftsmen built their expertise one job at a time, and your AI implementation should follow the same patient, and methodical approach.**

Advance: Applying AI in Manufacturing

Your shop floor is mapped, your data gaps are identified, and your craftsmen understand that AI is here to amplify their expertise, not replace it. Now comes the part where theory meets metal and plans meet production schedules. **Implementation in manufacturing looks different from other industries because downtime costs real money, and mistakes show up in scrapped parts.** The shops that succeed start small with one machine, one process, one product line, and prove value before expanding. They resist the temptation to flip every switch at once, no matter how impressive the vendor demonstration looked.

Here is what actually happens in the first ninety days: your team will find problems the software did not anticipate, your machinists will discover workarounds that work better than the manual suggests, and you will wonder at least twice whether you made the right decision. That is normal. **The difference between shops that succeed and shops that abandon their AI investment comes down to patience and repetition.** Treat implementation like breaking in a new machine. It takes time to learn its quirks, calibrate for your specific conditions, and integrate it into existing workflows. **The machinists who have been skeptical will become your best troubleshooters once they see the system catch something they would have missed.** And the ones who were enthusiastic will learn that AI handles tedious calculations so they can focus on the judgment calls that actually require twenty years of experience.

Three Best AI Programs for Precision Machining

Before you get excited about any AI vendor's promises, understand this: every single one of them will tell you implementation takes two weeks and integration is seamless. Plan for at least 3 months and get the vendor's definition of 'seamless' in writing.

1. *Autodesk Fusion 360 with AI Design Assistance* - The Complete Manufacturing Brain

What it does: AI-powered design optimization, material usage optimization for small production runs using smart systems that figure out how to use the least material while maintaining strength, automated nesting for sheet metal, and integration with computer-controlled equipment.

Why manufacturers choose it: Seamlessly integrates AI capabilities with existing CAD workflows without disrupting established design processes.

Real impact: *Metal Fabrication* cut material waste from 25% to 8% within six months.

Monthly cost: $200/month with manufacturing extensions.

Best for: Small to medium manufacturers focused on custom work and material optimization.

2. *MachineMetrics with AI Predictive Maintenance* - The Production Optimization Engine

What it does: AI-powered equipment monitoring, predictive maintenance alerts, production scheduling optimization, and comprehensive performance analytics.

Why precision manufacturers choose it: ***MachineMetrics*** prevents costly equipment failures while optimizing production schedules for maximum efficiency.

Real impact: The manufacturing company achieved a 45% reduction in unplanned downtime and a 23% improvement in product quality.

Monthly cost: Pricing varies greatly but generally runs about $200/month per machine.

Best for: Precision manufacturing shops with expensive equipment and tight tolerances.

3. *Mastercam with AI Tool Path Optimization* - The Programming Intelligence Engine

What it does: ***Mastercam*** is an advanced computer-controlled software solution that uses artificial intelligence to automate tool path optimization, enhance collision detection, and accurately predict machining times.

Why machinists choose it: The AI integrates conventional programming proficiency with machine learning to enhance manufacturing processes, resulting in increased speed, safety, and efficiency.

Real impact: Machine shops report a 35% reduction in programming time and a 28% improvement in surface finish quality.

Monthly cost: $350-$650, depending on implemented modules.

Best suited for: Machine shops specializing in complex parts with tight tolerances.

Cabinet Manufacturing Excellence

We often hear, "AI is going to put skilled tradesmen out of business." Really? Tell that to *Desco Professional Builders* in Ellington, Connecticut. Cabinet makers and woodworking shops represent some of the most skilled artisans in manufacturing, combining traditional techniques with precision requirements.

Success Story *Desco Manufacturing*

When the 2008 recession hit, construction work dried up, and most contractors were just trying to survive. What does *Desco* do? They doubled down to bring automation and cabinet production in-house. Sounds crazy, right?

With AI advancements, here is what happened next—and why it matters for anyone wondering whether AI enhances or replaces craftsmanship.

The Counterintuitive Move That Worked

Desco has always farmed out its cabinet work like most general contractors. But when projects became scarce, they saw something the "experts" missed: an opportunity to control their own destiny while helping other struggling contractors with the same supply chain headaches.

Shop supervisor Shane Bruscoe oversaw the introduction of ***Microvellum*** software, acknowledging automation's contribution to craftsmanship. This is not your basic design program. It is a system that connects computer-aided design directly to automated cutting and assembly workflows designed specifically for cabinet makers who know what they are doing.

The Smart Shop Layout (and Why Size Matters)

Desco* did not automate everything.** In their 5,000-square-foot automated section of their warehouse, ***Microvellum drives the precision work. Cutting, drilling, edge banding, and material optimization all involve mathematical calculations that are more efficiently and accurately performed by computers than by humans, regardless of experience level.

However, a visit to their 10,000-square-foot traditional warehouse reveals the true site of exceptional craftsmanship. Hand assembly, custom fitting, quality inspection, and the kind of problem-solving that comes from years of experience with wood, glue, and finicky customers.

The Numbers That Tell the Real Story

Prior to automation, 80% of internal work orders for *Desco's* construction projects **were initiated within the organization**, while 20% originated from external sales.

Following automation, internal work orders accounted for 20%, while **80% of external sales were generated by contractors throughout Connecticut**.

Think about that reversal for a minute. **Automation** did not just improve efficiency—it fl**ipped their entire business model and opened markets** *Desco* did not compete in before.

The Lesson Every Small Manufacturer Should Learn

As *Desco* President Anderson puts it, "If we had not automated when we did, **we would not have had the ability to compete against guys doing that work**. This has opened new doors that we had not played in."

Translation: a**utomation made *Desco* craftsmen more valuable, not less.** They could now deliver precision and speed that matched larger manufacturers—plus the personal service and customization that only a local shop can provide.

The irony? By implementing automation during the recent economic downturn, *Desco* developed competitive advantages that many shops without automation are still missing today.

Three Best AI Programs for Cabinet Manufacturing

1. *FinishMatch AI* - The Cabinet and Woodworking Specialist

What it does: AI analyzes thousands of wood finish combinations to predict precise color matches using visual prediction tools for customer decisions, integrated with traditional woodworking processes.

Why cabinet makers adopt it: *FinishMatch* addresses the specific challenge of matching wood stains and varnishes across different species and lighting conditions, eliminating material waste.

Real impact: Cabinet shops have experienced a 40% reduction in project completion time and a 67% increase in customer satisfaction.

Monthly cost: Custom pricing based on shop size.

Best suited for: Custom cabinet makers and woodworking shops that prioritize finish quality and customer satisfaction.

2. *Cabinet Vision with AI Optimization* - The Complete Cabinet Design Brain

What it does: AI-powered cabinet design, material optimization, cut list generation, and manufacturing workflow integration.

Why cabinet manufacturers choose it: Provides end-to-end design and manufacturing optimization specifically for cabinet production.

Real impact: Cabinet shops report a 30% reduction in material waste and a 45% increase in design-to-production workflow speed.

Monthly cost: $200-$400/month, depending on features.

Best for: Cabinet manufacturers focused on efficiency and customer design collaboration.

3. *Thermwood CNC* with AI Programming - The Cabinet Manufacturing Automation Engine

What it does: AI-enhanced computer-controlled programming for cabinet components, automated nesting, quality control monitoring, and production scheduling.

Why cabinet manufacturers choose it: ***Thermwood*** combines traditional woodworking expertise with AI automation for consistent, high-quality production.

Demonstrated impact: Cabinet manufacturers report a 50% decrease in programming time and a 35% increase in production consistency.

Monthly cost: Varies based on equipment configuration.

Best suited for: Cabinet manufacturers looking to automate production without sacrificing quality.

Sheet Metal Fabrication Excellence

Success Story
Custom Fabrication

Each HVAC project required different configurations, making manual planning nearly impossible. A typical commercial job might require dozens of different duct sizes, angles, and fittings—all cut from standard sheet metal stock. Without sophisticated nesting software, the fabrication shop was losing 25-30% of its material to waste and spending 4-6 hours per job just planning cuts and coordinating production.

The AI-Enhanced Solution

They implemented ***Autodesk Fusion 360*** with manufacturing extensions specifically designed for sheet metal fabrication. The key was not just the software; it was the AI-powered nesting algorithms that could optimize complex cutting patterns for small batch production.

Implementation Strategy:

- ✓ ***Fusion 360*** offers sheet metal topology optimization with manufacturing extensions.
- ✓ AI-powered **nesting algorithms** for complex custom work.
- ✓ Smooth **integration with current plasma cutting systems.**
- ✓ **Professional Development Program** for Shop Foremen and Lead Fabricators.

Results after 6 months:

- Material waste was reduced from 25–30% to 8–12%.
- Job planning time decreased significantly, from 4–6 hours to just 45 minutes.
- Setup time improved by 38% through the implementation of optimized cutting sequences.

- Profit margins increased by 28% as a result of greater material efficiency.
- The company's reputation has been strengthened due to precise estimates and consistent delivery.

Three Best AI Programs for Metal Fabrication

1. *SigmaNEST with AI Nesting* - The Sheet Metal Optimization Master

What it does: AI-based nesting optimization, analysis of material usage, cutting path refinement, and remnant management.

Why fabricators choose it: ***SigmaNEST*** maximizes material usage while minimizing cutting time and tool wear.

Real impact: Metal fabricators have observed a 25-40% increase in material utilization and a 30% decrease in cutting time.

Monthly cost: $300-$600/month based on features.

Best for: Sheet metal fabricators focused on material efficiency and cost reduction.

2. *FabriWIN with AI Scheduling* - The Fabrication Workflow Intelligence Engine

What it does: AI-enhanced production scheduling, machine capacity optimization, job routing, and delivery coordination.

Why fabrication shops choose it: Optimizes complex multi-machine workflows while maintaining quality and delivery commitments.

Real impact: Fabrication shops record a 35% increase in on-time delivery rates and a 25% rise in machine utilization.

Monthly cost: $200-$500/month, depending on shop size.

Best for: Multi-machine fabrication shops with complex scheduling requirements.

3. *Plainsight with Vision AI* - The Fabrication Quality Assurance System

What it does: ***Plainsight*** uses computer vision and machine learning to automate quality inspection on production lines. The system detects defects, foreign objects, and anomalies that manual inspection often misses. Cameras enabled with their vision AI can identify discoloration, measurement variations, surface imperfections, and other quality issues in real time, triggering alerts or automatic sorting before defective parts move downstream.

Why quality-focused fabricators choose it: Manual quality inspection catches roughly 80% of defects on a good day. Fatigue, distraction, and the sheer volume of parts make perfection impossible for human inspectors. ***Plainsight*** works around the clock without getting tired, and the machine learning improves over time as it sees more examples of what good and bad parts look like.

Real impact: Manufacturers using vision AI for quality inspection report catching defects earlier in the production process, reducing costly rework and scrap. The system can reduce downtime by 36% compared to traditional manual approaches by identifying problems before they cascade into bigger issues.

Cost: They offer tiered plans from startup-friendly entry level to enterprise $49/replica/year up to $189/replica/year; based on the number of "filters" and deployment scale.

Best for: Metal fabricators who need consistent quality inspection across high volumes, particularly shops where defects are costly—either in customer relationships, rework time, or material waste.

Connect: Making AI and Craftsmanship Work Together

You get better results when AI does the math, and your craftsmen make the judgment calls. This collaboration leads to production capabilities that neither one could achieve alone.

Materials Optimization Integration

AI Handles: The system generated efficient nesting patterns for complex sheet metal parts, minimizing both material waste and production time.

People Provide: Expert fabricators reviewed AI recommendations, adjusted for material characteristics and tool limitations, and optimized cutting sequence based on 18 years of experience.

Working Together Results: A 42% reduction in material waste with faster production timelines.

Quality Control Integration

AI Handles: Continuous measurement and analysis of part dimensions, surface finish, and tolerance adherence.

People Provide: Experienced machinists interpreted AI data within the context of customer requirements, making adjustment decisions based on application knowledge.

Working Together Results: Zero quality issues, with a 34% faster inspection process and increased customer satisfaction.

What Could Go Wrong (And How to Prepare)

Implementing AI in manufacturing seems ideal in sales demos and case studies, but the reality is often different. Then reality hits.

When the AI system crashes during a critical job:

It will happen. Your backup plan cannot be "call tech support and wait three hours." Have manual procedures documented for every critical operation. Your experienced craftsmen need to know how to complete jobs without the AI, even if it takes longer. **Test these backup procedures before you need them.**

When predictive maintenance gives false alarms:

The system will cry wolf. Maybe the sensor is dirty. Maybe the algorithm is oversensitive. Your team will want to ignore all alerts after the third false positive. Establish clear protocols for validating AI warnings before either acting on them or dismissing them. Track false positives and work with your vendor to tune sensitivity levels.

When material optimization creates cutting patterns that work on paper but fail in practice:

The AI does not understand that some materials behave differently at certain temperatures, or that your specific plasma cutter has quirks the algorithm never learned. **Your craftsmen need authority to override AI recommendations** when their experience tells them something is wrong. Document these overrides and feed them back to improve the system.

The learning curve nobody mentions:

Months three through six are the danger zone. The novelty has worn off. Your craftsmen are frustrated because the system does not work as smoothly as promised. **Productivity might actually be down compared to your old methods.** This is when shops quit.

Push through. Schedule weekly check-ins focused on specific problems, not general complaints. Celebrate small wins. Track improvements that matter to individual craftsmen, not just overall shop metrics. **Give your most skeptical worker veto**

power over the implementation timeline. When they finally say it is working, everyone else will believe it.

Keep: Maintaining Manufacturing Excellence

Manufacturing companies that successfully adopt AI **develop advantages that grow over time** by building systems that improve craftsmanship skills while maintaining the expertise culture that defines their value.

Scaling What Works

Review your first six months of data. **Which AI insights led to your biggest improvements in quality or efficiency?** Double down on those areas. If AI-assisted programming reduced setup time by 35% and improved surface finish quality by 28%, expand that system to cover more complex operations. If predictive maintenance prevented three equipment failures and saved $18,000 in repair costs, implement monitoring across all critical machines.

Many manufacturers make the mistake of adding new AI tools instead of optimizing the ones they already have. **Resist this temptation.** Perfect your current systems before expanding.

Avoiding the Success Trap

Success can create overconfidence for manufacturers. **Do not start automating everything just because your first AI tools worked well.** Customers still value the craftsmanship and problem-solving expertise that only experienced, skilled workers can provide. **AI should enhance these capabilities, not replace them.**

Monitor both productivity metrics and customer satisfaction scores closely. If quality standards start declining or customer relationships become impersonal, investigate whether technology is interfering with the craftsmanship that built your reputation.

Marten Machining's 30 Year Evolution

Al Marten's approach to continuous learning has created a sustainable model for AI-enhanced manufacturing. He developed an:

Ongoing Learning Culture: "You can always learn," Marten says. "After all my years in the business, I am still learning." This mindset extends to AI implementation; each innovative technology is approached with curiosity rather than fear.

Apprenticeship and Training: *Marten Machining* has trained dozens of future machinists, combining traditional craftsmanship skills with modern AI-enhanced CNC capabilities. They volunteer at local high school machine shop programs and partner with regional tech schools.

Strategic Technology Investments: Instead of investing in the latest technology, Marten prioritizes tools that enhance human expertise. Specific customer needs that require capabilities beyond what manual methods can provide drive each CNC investment.

Customer Partnership: "What has driven our growth from the beginning is our ongoing commitment to invest in resources to produce specialty, high-tolerance parts," Marten explains. AI enhancement advances this mission without altering it.

Industry Results and Competitive Advantages

Cabinet Manufacturing Industry Transformation:

AI finish-matching tools **reduce customer decision time** while preserving the impact of craftsman consultations.

- Computer-controlled programming assistance allows **more focus on design creativity and customer service.**
- Automated nesting **optimizes material usage** while maintaining traditional joinery and fabrication quality.

Precision Machining Evolution:

AI-assisted programming reduces setup time, **creating more capacity for complex and quality work**.

- Predictive maintenance prevents failures that disrupt manufacturing schedules.
- **Quality monitoring** helps maintain the tight tolerances that define precision manufacturing.

Small Shop Competitive Advantages:

AI tools level the playing field, letting you compete with larger manufacturers without giving up personal service.

- Automation handles routine tasks, **freeing craftsmen** for complex problem-solving and complicated manufacturing.

- **Technology investments improve** what the business does and help it stand out as a premium option.

The Manufacturing Advantage

The most successful small manufacturers **do not treat AI as a replacement technology**. They utilize it as an advanced tool that requires skilled operators to fully realize its potential.

Key Success Elements:

- AI **handles data analysis** and routine calculations.

- **Craftsmen provide expertise,** problem-solving, and quality judgment.

- Customer **relationships remain personal** while service capabilities grow.

- **Apprenticeship programs** teach both traditional skills and AI tool mastery.

Manufacturing excellence is not about choosing between traditional craftsmanship and modern technology. It is about using **AI to make skilled workers more capable,** valuable, and competitive than before.

Small manufacturers thoughtfully using AI create advantages that larger competitors cannot easily reproduce. The combination of advanced technology, personal service, and craftsmanship takes years to develop.

Here is what should keep manufacturers who wait to implement AI up at night: While you are deciding whether AI is right for your shop, your competitor three towns over just cut their lead times by 40% and their material waste by half. Their craftsmen are landing jobs; yours have to turn away. Their customers are bragging about the precision and speed that seemed impossible two years ago.

How long before your customers notice?

Chapter 14

Agriculture

Growing Your Operation With AI

Livestock, crops, and **grapes** do not concern themselves with technology hype. If you are a **farmer, rancher,** or **vintner,** you make decisions based on what improves yields, reduces costs, and strengthens the long-term health of your land and animals. Your business is built on cycles, seasons, margins, and experience. AI becomes valuable only when it supports those realities in practical ways.

AI in agriculture succeeds when it enhances the good decisions you already make, reveals conditions you cannot see from the ground, or alerts you early enough to prevent costly problems. **When implemented thoughtfully, it becomes a tool that strengthens your operation, protects your margins**, and helps ensure that future generations can continue the work you have built.

This chapter covers three distinct agricultural operations that benefit from different AI approaches. **Crop management** for improved planting decisions, irrigation efficiency, and pest control. **Livestock observation** to reduce veterinary costs and improve herd health. **Vineyard-level analysis** to strengthen product quality and margins.

Success Story
Heritage Farm

Heritage Farm in central Illinois has been family-owned and operated for three generations. They farm 1,800 acres of corn and soybeans. The fields had productivity variations, but they managed them by applying uniform fertilizer across each field. This was the standard practice, which was wasteful.

In 2022, the operation started using ***OneSoil,*** which revealed patterns they had suspected but never proved, revealing certain areas consistently underperformed due to soil differences and drainage issues. Previously underperforming areas saw increased yields after receiving the appropriate nutrient blend.

Results After 1 Year: 12% reduction in fertilizer costs by targeting applications to zones that actually needed them. They saved $21,000 in fertilizer costs alone while **increasing an average yield by 7 bushels an acre**.

Plantix was added the next season after finding what he thought was fungal rust in a 40-acre corn section. It identified the problem as a nutrient deficiency instead, specifically a sulfur shortage. A $300 sulfur application fixed the problem. (A fungicide treatment for rust would have cost $1,800 and would not have solved anything).

Results After 3 Years: The farm operates with 15% lower costs and 8% higher yields compared to their baseline.

Agricultural AI is not about replacing knowledge; it is about providing knowledge that was once thought to be only known by time and nature. **The best implementations combine generations of hands-on experience with AI systems that monitor, analyze, and alert in ways human observation cannot match**. AI helps to see patterns you would miss and catch problems before they become expensive disasters.

The Agriculture Transformation Challenge

Agriculture presents unique AI adoption challenges because true expertise is earned through years of direct experience. Weather patterns, soil characteristics, microclimates, and livestock behavior cannot be fully understood through theory alone. AI must be trained on what makes your operation distinct.

At the same time, buyers expect consistency, traceability, and documented quality. Producers who rely solely on manual record-keeping or visual inspection are increasingly disadvantaged. **The right AI system creates stronger control over variability while keeping your judgment at the center of every decision.**

Agricultural excellence is not a choice between tradition and technology. It is the strategic combination of both. Operators who use AI thoughtfully create advantages that are extremely difficult for others to replicate.

Trust: Building Confidence Before Implementation

Agricultural AI adoption starts with demonstration, not faith. You and your team need to see how the technology actually works before you will trust it enough to use it consistently. **Start small with a proof-of-concept on a limited acreage or a subset of animals.** Work that across one season before expanding to your entire operation. Ensure accuracy before expanding. Compare AI recommendations against your current practices.

Expect it to be 3-4 months before your team fully trusts the system. Agricultural experience is earned over seasons, and you need to see AI recommendations proven correct across different weather conditions and challenges before you rely on it for important decisions.

Ready: Preparing Your Operation

Farmers are the original skeptics, and for good reason. You have watched enough miracle solutions come and go; the supplement that was supposed to double yields, the equipment that broke down during harvest, the consultant who had never actually operated a farm telling you how to run yours. So when someone shows up talking about AI and satellite imagery and predictive analytics, your instinct is to nod politely and wait for them to leave. We get it. **But this time, the technology actually works—if your operation is ready for it.**

Here is what the vendors conveniently forget to mention: **agricultural AI needs connectivity, and most farms do not have it where it matters.** That soil moisture sensor is useless if it cannot transmit data from the back forty. That livestock-monitoring camera will not do you any good if your barn sits in a cellular dead zone. Before you buy any AI system, drive your property with your phone and map where you actually have signal. Then get honest about whether the coverage matches where you need monitoring. Rural internet has improved, but **"improved" and "reliable" are not the same thing.** There is no shortage of operators who spent thousands on precision agriculture equipment only to discover their fields had connectivity gaps that turned real-time monitoring into next-day monitoring, which defeats the entire purpose.

Your data situation matters even more than your connectivity. AI systems learn from historical information, and they need baselines to detect problems. What is normal soil moisture for your south field in July? What does healthy vegetation look like on satellite imagery for your specific crops? How much does your herd's movement pattern typically vary day to day? If you do not have answers to these questions documented, **the AI will spend its first season learning what you already know instead of catching what you would miss.** That is not wasted time; you just need to plan for it. Expect three to six months of baseline collection before any agricultural AI starts delivering the insights the sales brochure promised.

Then there is the equipment question. Precision agriculture tools need to talk to your existing machinery, and farm equipment manufacturers have not exactly prioritized interoperability. That variable-rate planter controller might work beautifully with the AI prescription maps, or it might require a $4,000 retrofit and a dealer visit that takes three weeks during your busiest season. **Get specific compatibility answers in writing before you commit. "Should work with most major brands" is not an answer; it is a warning.**

Talk to your crew before you install anything. The farmhand who has worked your land for fifteen years notices things that satellites cannot see: the way water pools in that one corner after rain, the section where deer pressure is worst, the equipment that runs hot when it is about to have problems. **AI should amplify that knowledge, not ignore it.** If your team thinks the new technology is there to replace their judgment or monitor their productivity, they will resist it. If they understand it is there to catch the things they cannot be everywhere to see, they will become your best source of feedback on what is working and what is not.

Finally, **pick your starting point based on your biggest pain, not the flashiest technology.** If you are losing animals to illnesses, you catch too late, livestock monitoring makes sense. If your input costs are eating your margins because you are over-applying across variable soil, precision application AI is your priority. If you are a vineyard losing fruit quality to microclimates you cannot track manually, start there. **The operations that succeed with agricultural AI are the ones that solve real problems first and expand from proven results not the ones that try to digitize everything at once.**

Your land, your animals, and your crops do not care about technology trends. **They respond to good decisions made at the right time. AI just helps you make more of those decisions, faster, with information you could not gather on your own.**

Baseline Requirements

Infrastructure Needs

- **Reliable cloud-based systems** are essential for consistent monitoring and data management.
- While satellite internet can be used, **it may cause delays** in monitoring activities.
- Implementing battery backup systems is important to prevent data loss during power outages.
- **Ensure that alerts can be received** and data viewed **in the field** for timely decision-making.

Team Preparation

- Designate at least **one person** in your operation **who is capable of troubleshooting** sensor issues and interpreting data alerts.
- **Provide clear descriptions of what the AI system monitors,** such as stating "camera watches cow gait patterns to catch lameness early" instead of generic statements like "AI analyzes movement data."
- **Clarify** for the team **what actions are required** for different types of alerts generated by the system.

Data Baseline Requirements

- **Establish normal** vegetation **patterns** to serve as reference points for detecting anomalies.
- **Collect sufficient baseline data** before relying on alerts to ensure their reliability.
- **Calibrate AI recommendations** based on these established baselines.

What Could Go Wrong

Skipping the baseline period means false alerts and wasted time investigating non-issues. You will lose team confidence in the system before it ever proves valuable.

Crop Management: The Reality Behind Growing Technology

Precision agriculture sounds expensive and complicated. Satellite monitoring, variable rate application, predictive analytics---the terminology alone can make a farmer stick with what you know. **There is no denying that ag was one of the first real winners in the AI game.**

Here is what actually happens when crop operations implement AI. You discover which parts of your fields consistently underperform and why. You catch pest infestations or nutrient deficiencies three weeks before they would show symptoms visible to the eye. You reduce pesticide applications by 35-40% because you only treat the 20% of acreage that actually needs it. **Your input costs drop while yields improve.**

Top Three AI Tools for Crop Management

Crop management AI tools represent different entry points based on your operation size and technical comfort level. All three use satellite imagery and machine learning. Where they differ is in complexity, cost, and how much control you want over the data.

1. Plantix - Pest and Disease Detection

What It Does: ***Plantix*** turns your smartphone into a crop disease diagnostic tool. Take a photo of a sick plant, and deep learning algorithms compare it against a database of 850+ crop diseases, pests, and nutrient deficiencies. You get a diagnosis in seconds with treatment recommendations.

The Cost: Free to use with optional premium features. More than 10 million downloads make it the number one agricultural app for damage detection and pest control worldwide.

Best For: Small farms needing immediate problem diagnosis. Particularly valuable for specialty crop growers managing multiple varieties where disease identification expertise may be limited. Works for vegetables, fruits, grains, and most commercial crops.

What Could Go Wrong: Image quality matters. Blurry photos or poor lighting reduce accuracy. The AI is trained on global crop diseases, so regional variations might not always match perfectly. Use it as a diagnostic starting point, not the final word especially for expensive treatment decisions.

2. *OneSoil* - Field-Level Precision Agriculture

What It Does: ***OneSoil*** uses satellite imagery and neural networks to divide your fields into productivity zones. The system tracks crop health throughout the growing season, generates variable rate application maps for fertilizer and lime, and provides yield forecasting before harvest. All without requiring any hardware installation.

Hardware You Need: Nothing. The system uses publicly available satellite data from Sentinel-2 and Landsat. You access everything through a web browser or mobile app. If you want to implement variable rate application, you need equipment that can read prescription maps---but the AI analysis itself requires zero hardware.

The Cost: Freemium model. Basic field monitoring is free. Pro subscription is variable but averages $29/month and adds variable rate maps, productivity zone analysis, and historical field data.

Best For: Mid-sized operations (100-2,000 acres) ready to move beyond uniform field treatment. Particularly valuable for farms with variable soil types or irregular field shapes where blanket application wastes inputs.

What Could Go Wrong: Satellite imagery depends on clear weather. Cloud cover during critical growth stages can leave gaps in monitoring. The system works best with crops that show clear spectral differences when stressed. Row crops like corn and soybeans are ideal, but some specialty crops may not generate useful data.

3. *EOSDA Crop Monitoring* - Enterprise-Grade Field Intelligence

What It Does: ***EOSDA Crop Monitoring*** combines satellite imagery, weather data, and machine learning to provide field-level vegetation monitoring, productivity analytics, and risk assessment. The system tracks 14 different vegetation indices, monitors soil moisture, and generates alerts for crop stress events. More sophisticated than ***OneSoil,*** with more data sources and analysis depth.

Tools You Need: Computer and internet access. The platform uses satellite data from multiple sources, including Sentinel, Landsat, and Planet. All analysis happens in the cloud.

The Cost: Free Plan or Paid subscription starting at $101 per month, covering up to 1,000 hectares (2,471 acres). Pricing scales with acreage.

Best For: Larger operations (1,000+ acres) or mid-sized farms willing to invest in comprehensive monitoring. Operations managing multiple crop types across diverse field conditions get the most value from the detailed analytics.

What Could Go Wrong: The depth of data can be overwhelming if you are not ready for it. You get 14 vegetation indices when you probably only need 3-4 for most decision-making. The learning curve is steeper than simpler tools, and the monthly cost adds up. Make sure you will actually use the data before committing to a year-long subscription.

Livestock Health Monitoring

The Reality Behind the Technology

Livestock health monitoring sounds like something only industrial dairy operations can afford. Camera systems, collar sensors, and computer vision leave the impression of high-tech equipment that costs more than many farmers spend on tractors.

When livestock operations implement AI, you catch illnesses 2-3 weeks before animals show visible symptoms. Veterinary costs drop 30-35%, because you treat problems early instead of managing full-blown infections.

Production improves because healthier animals gain weight faster, produce more milk, and have better reproductive success.

Success Story
Bles Dairy

At *Bles Dairies Farm* in the Netherlands, the cows wear sensor collars connected to ***Connecterra IDA***, short for Intelligent Dairy Assistant. The system, built on Google's TensorFlow machine learning platform, tracks seven behavioral patterns: eating, drinking, ruminating, walking, lying, standing, and activity levels. **The AI learns what normal looks like** for each individual cow, then flags when behavior deviates from that baseline.

The practical application is straightforward. Sales Director Leonie ten Dam explains: "The system informs the farmer about the situation of a cow. For example, the cow is eating more but ruminates a lot less." That kind of insight, delivered automatically to a smartphone, lets farmers intervene before small problems become expensive ones.

Top Three AI Tools for Livestock Health Monitoring

1. *CattleEye* - Early Lameness Detection

What It Does: CattleEye uses a standard security camera mounted above your milking parlor exit to monitor cow movement and identify lameness up to 23 days before you would notice it yourself. No collars, no wearables, no tagging animals. Just a camera watching cows walk.

Hardware You Need: A basic security camera costs around $150 and mounts above your dairy building exit. That is it. ***CattleEye*** connects to your system through the cloud and starts analyzing cow movement patterns immediately.

The Cost: $1.45/month per cow. Operations save between $320 and - $420 per cow annually by catching early lameness. At Highfields Farm in Cheshire with 1,050+ Holstein cows, they achieved an 8% reduction in lameness in just 18 months. Their vet could not find a single lame cow during an audit.

Best For: Dairy operations of any size. The system monitors over 100,000 cows worldwide without requiring expensive hardware on each animal. If you have cows walking through a milking parlor, this works.

What Could Go Wrong: You need decent internet connectivity for cloud processing. Rural areas with spotty connections will struggle. The camera placement matters---9 feet in the air with a proper angle, and good lighting. Get it wrong and the system cannot accurately analyze gait patterns. Winter conditions with snow buildup can temporarily disable monitoring if cameras are not properly protected.

2. *SmaXtec* - Internal Health Monitoring

What It Does: ***SmaXtec*** uses a bolus (a pill-sized sensor) that sits in the cow's rumen and continuously monitors internal body temperature, activity patterns, and rumination behavior. The system detects health issues like metabolic disorders, calving complications, and heat stress days before external symptoms appear.

Hardware You Need: The bolus sensors for each animal plus base stations around your facility to collect data. The boluses are administered like any other pill---the cow swallows it, and it stays in the rumen, monitored continuously. You also need internet connectivity for the cloud-based analytics platform.

The Cost: Each bolus costs approximately $220- and lasts 4-5 years. System subscription runs around $5 per month per cow.

The Results: At Fair Oaks Dairy in Indiana, with 800 cows, they reduced treatment costs by 35% and improved early disease detection by 78% within the first year.

Best For: Medium to large dairy operations (200+ cows) where individual animal health tracking justifies the per-cow investment. Particularly valuable for high-value breeding stock where reproductive health monitoring prevents expensive problems.

What Could Go Wrong: The upfront investment adds up quickly when you are equipping hundreds of animals. Boluses can occasionally fail or be regurgitated, requiring replacement. The system generates a lot of data that requires interpretation---you need someone monitoring the platform daily and understanding what the alerts actually mean.

3. Connecterra Ida - Comprehensive Behavior Analysis

What It Does: ***Connecterra Ida*** uses neck collars with motion sensors to track detailed activity patterns---eating, ruminating, standing, lying down, and heat cycles. Machine learning algorithms establish baseline behavior for each animal and alert when patterns deviate, indicating potential health or reproductive issues.

Hardware You Need: A collar for each monitored animal plus gateway devices around your facilities to collect sensor data. The collars are rechargeable and designed to withstand typical farm conditions. An internet connection is required for the cloud platform to perform.

The Cost: The collar costs approximately **$200** per cow, with a monthly subscription of averaging **$3** per cow.

Best For: Operations wanting comprehensive health and reproductive monitoring without internal sensors. Particularly valuable for farms where heat detection and breeding efficiency directly impact profitability. Works for dairy cattle, beef cattle, and even sheep or goats with appropriate collar sizing.

What Could Go Wrong: Collars can be damaged or lost, requiring replacement. Battery management means regular recharging on a rotation basis. The system works best when you have baseline behavior data for each animal, so there is a learning period of several weeks before alerts become reliable. Animals that frequently rub against fences or equipment may damage collars more quickly.

Vineyard & Winery Management

The Reality Behind the Technology

In the vineyard, you face the same precision agriculture challenges as any crop operation, but with a higher per-acre value and narrower quality windows. In the winery, you manage fermentation processes where small temperature or timing variations affect the final product's market value by thousands of dollars per barrel. **If you implement AI, you catch vineyard stress days before they affect grape quality. You optimize irrigation to maximize flavor concentration without wasting water.** In production, you monitor fermentation with precision, eliminating batches lost to temperature variations or timing errors. Your consistency improves, meaning higher-quality wines and better margins.

Success Story
Gamble Estates Winery

Tom Gamble brings more than 100 years of family farming history to his 170 acres in Napa Valley. His father farmed grapes. His grandfather farmed grapes. So, when **autonomous tractors with AI sensors became available,** you might expect a third-generation farmer to be skeptical. Instead, Gamble became an early adopter, purchasing a ***Monarch AI Tractor***.

The tractor is, "basically a computer on wheels." As it drives each row, its AI sensors collect data on vine health, soil conditions, and crop status. The **machine learning algorithms process this information** to help Gamble make what he calls "precision farming" decisions. The system learns the vineyard over time, building an increasingly accurate map of conditions across his 170 acres.

The results go beyond convenience. Gamble estimates the electric tractor saves about 40 to 50 tons of carbon emissions annually, which matters for regulatory compliance and increasingly for consumer perception. But he is clear-eyed about what AI can and cannot do.

"It is not going to completely replace the human element of putting your boot into the vineyard, and that is one of my favorite things to do," he says. "But **it is going to allow you to work more smartly, more intelligently, and in the end, make better decisions under less fatigue."**

Farming is physically exhausting work. AI that reduces fatigue is not about replacing farmers. It is about keeping them sharp enough to make good decisions when those decisions matter most.

Top Three AI Tools for Vineyard & Winery Management

1. *VineView* - Precision Viticulture Monitoring

What It Does: ***VineView*** uses satellite imagery and machine learning to monitor vineyard health, detect water stress, and identify disease pressure before visible symptoms appear. The system generates variable rate irrigation maps and provides harvest timing recommendations based on grape maturity patterns across different vineyard blocks.

The Cost: Subscription pricing starts around $12 per acre annually.

Best For: Vineyards with 20+ acres where within-vineyard variability affects quality. Particularly valuable for premium wine producers where grape quality directly determines bottle price. Works for any grape variety, though it is most proven with Cabernet Sauvignon, Pinot Noir, and Chardonnay.

The Results: Vineyards report a 15-25% water savings from optimized irrigation and quality improvements that command premium pricing. One Napa Valley producer documented $45,000 in increased revenue from a single vintage by harvesting different blocks at optimal ripeness rather than all at once.

What Could Go Wrong: Satellite-based monitoring depends on clear weather for accurate readings. Mountainous terrain or very small vineyard blocks may not generate useful data. The system works best when combined with ground-truthing---using the satellite data to identify areas needing attention, then verifying with physical inspection.

2. *Blended* with Barry AI Assistant - Cellar Operations Intelligence

What it does: ***Blended*** manages winery production from crush to bottle, with **Barry**, their AI assistant built specifically for wine production. Barry creates work orders automatically, answers questions about your cellar operations, and logs activity in seconds. The platform tracks fermentation, barrel inventory, blending trials, and TTB compliance documentation in one system.

Why winemakers choose it: The AI assistant handles the tedious documentation that pulls winemakers away from actual winemaking. Instead of spending harvest weekends catching up on paperwork, you ask Barry to create work orders and track what is happening across multiple tanks simultaneously.

Real impact: Winemakers report significant time savings during harvest, with one cellar manager noting she no longer works days off just to keep up with documentation. The system reduces mistakes in barrel tracking, blend management, and work order accuracy.

Monthly cost: Pricing varies depending on the winery size but may run in the $500 a month range for a small winery.

Best for: Wineries wanting AI-powered production management that adapts to their workflow rather than forcing rigid structures. Works particularly well for operations tired of spreadsheets but not ready for enterprise-level complexity.

3. VineSpring - Direct Sales Intelligence

What It Does: ***VineSpring*** combines customer relationship management with AI-powered recommendations for tasting room operations and wine club management. The system analyzes purchase history, tasting notes, and customer preferences to suggest wines during tastings and predict which club members are likely to cancel. It also optimizes inventory allocation between tasting room, club, and wholesale channels based on profitability.

Hardware You Need: Point-of-sale system integration (works with most major wine POS systems like Commerce7 and WineDirect). Tablets or computers for tasting room staff to access customer profiles and recommendations during tastings.

The Cost: Subscription fees vary from **$300-500** monthly depending on transaction volume.

Best For: Wineries with significant direct-to-consumer sales (tasting room and wine club). Most valuable for operations doing $500,000+ in annual DTC revenue where customer intelligence directly impacts profitability. Less useful for production-focused wineries selling primarily through distribution.

The Results: Bella Luna Winery in Oregon increased average tasting room purchases from $87 to $124 per customer using the AI recommendations and reduced their wine club dropouts from 22% a month down to 12% using predictive retention alerts.

What Could Go Wrong: The system is only as good as your data quality. Incomplete customer profiles or inconsistent purchase recording reduce recommendation accuracy. Tasting room staff need training to use the system effectively rather than ignoring the suggestions. Some customers find data-driven recommendations impersonal if not presented naturally.

Advance: Implementing and Optimizing

Implementation moves in stages: monitoring first, then action, then full integration. Rush this and you create chaos. **Move deliberately and you build capability.**

Implementation Timeline: Stages of AI Integration

First 90 Days – Monitoring and Validation

During the initial three months, **focus on continuous monitoring and validating the AI system's outputs.** Pay close attention to any overlooked problems or irregularities and make necessary adjustments during this time. Confirming the system's reliability now is essential for future success.

Months 4–6 – Action Integration

Next, **apply AI-driven recommendations directly.** This includes responding to system-generated alerts and starting to apply recommendations, such as targeted use of pesticides if utilizing precision agriculture tools. The emphasis should be on taking informed actions guided by validated data.

Months 7–12 – Full Operational Integration

Aim to have the AI system fully integrated into day-to-day activities by the latter part of the year. This involves **combining AI-generated information,** like weather forecasts, with human expertise to ensure a seamless exchange of knowledge. The goal is to create a cohesive workflow where technology and field experience complement each other.

Expected Outcomes and Results

- Earlier and **more accurate problem detection**, allowing for timely intervention.
- Enhanced overall **management through the integration of AI insights and your own judgment.**

Potential Risks: Scaling up operations before demonstrating a return on investment with a smaller segment **can result in costly failures across the organization.** Validate the approach works for your specific conditions before full implementation.

Connect: Integrating AI with Human Expertise

Agricultural AI works best when it **amplifies human expertise rather than replacing it.** The most successful operations use AI for what it does well---continuous monitoring, pattern recognition, early detection---while relying on people for judgment, context, and relationship management.

The Effective Division of Responsibilities

AI handles:

- 24/7 monitoring across acres or hundreds of acres or animals.
- **Detecting subtle changes humans would miss.**
- Identifying patterns across multiple growing seasons or production cycles.
- **Alerting to problems** at the earliest possible stage.

Operators provide:

- Experience in field history, soil conditions, and local weather patterns.
- Judgment about whether **AI recommendations make sense given current conditions.**
- Physical inspection when alerts require verification.
- **Relationships with buyers, vets, agronomists, and suppliers.**

Working together results:

- Problems are caught at the **stage where solutions are still cheap** and effective.
- **Resources** get applied where they can **have an impact** uniformly across the operation.
- Consistency in quality improves, as **differences are spotted and resolved** at an early stage.
- Because operators have access to thorough information, they are able to **make quicker and more informed decisions.**

Example from livestock operations: AI sensors detect a cow with elevated body temperature and reduced rumination at 2 AM. The system alerts the operator, and by morning, the cow is physically examined. Based on years of operator experience, an early-stage metastasis is diagnosed and treated immediately. The combination of **early AI detection and operator expertise** allows a serious infection to be identified that would have required extensive treatment and caused significant milk production loss.

Example from crop operations: Satellite imagery shows a pattern of vegetation stress across a 20-acre section. The farmer recognizes that the section has shallow topsoil over hardpan and adjusts irrigation accordingly. The **AI identified the problem; the farmer's knowledge of the field explained why it occurred** and expertly determined the solution.

Keep: Maintain and Evolve Your System

Agricultural AI is not a "set it and forget it" technology. The **system learns and improves as it accumulates more data** about your specific operation. Your role after 6-12 months shifts from implementation to optimization and expansion.

Maintaining and Optimizing Your Agricultural AI System

System Maintenance Priorities

- Ensure that all hardware is installed correctly and secured, with particular attention after adverse weather events to maintain system integrity and reliability.
- Regularly **seek opportunities to improve accuracy and add new features that enhance system capabilities** and operational outcomes.

Performance Tracking

- Monitor the system's **effectiveness in detecting issues and contributing to improved yields** or production levels.
- Track how the system expands access to better buyers, **supporting stronger business relationships** and market positioning.
- Focus on the **system's ability to deliver insights** and benefits that surpass traditional monitoring methods.

Expansion Opportunities After the First Year

- Explore possibilities to **grow and scale your AI systems.**
- **Adapt to evolving operational needs** and market opportunities.

Building Competitive Advantage

- By delivering reliable and uniform results, you can **build a strong reputation** for excellence in your products and services.
- Leverage **AI-driven insights to implement advanced management practices** that set your operation apart.
- **Attract and retain talented workers** and secure valuable partnerships by demonstrating industry leadership and technological advancement.

Team Development

- **Foster a culture of development and learning** within your team to support ongoing growth and adaptation to new technologies.

The Agriculture Advantage

The most successful agricultural operators do not treat AI as a replacement for the land knowledge and farming instincts they have built over generations or a betrayal of the traditions that define their operation. They use it as a **farm management amplifier that requires experienced operators** to fully realize its potential.

Key Success Elements:

- **AI handles soil monitoring, livestock health tracking, yield prediction, and resource optimization.**
- Farmers provide agronomic judgment, animal husbandry expertise, weather interpretation, and the land stewardship that produces quality year after year.
- **Buyer relationships remain personal** while operational capabilities expand to meet premium market demands.
- Quality stays consistent because **AI catches problems before they become crop failures or herd health crises.**

Agricultural excellence is not about choosing between traditional farming wisdom and modern technology. **It is about using AI to make experienced operators more informed, more efficient, and more resilient than ever before.**

Agricultural operations thoughtfully using AI create advantages that large industrial farms cannot easily replicate. **The combination of precision monitoring, sustainable practices, and genuine expertise in your specific land and animals takes generations to develop.** A corporate farm is just acreage. A farmer who knows how the south field drains differently after heavy rain, who recognizes when the herd's behavior signals something wrong before symptoms appear, who understands the microclimates that make one vineyard block produce differently than the next? That is agricultural expertise no satellite can replace.

The rancher who implemented AI livestock monitoring did not lose sight of his herd's health. **He stopped losing calves to illnesses he caught too late** and started intervening when the data showed problems his daily rounds would have missed. The vineyard that added precision irrigation did not sacrifice its land. They stopped over-watering the blocks that did not need it and started producing the consistency that premium buyers demand. The row crop operation that deployed variable-rate application did not abandon good agronomy. They stopped blanket-spraying fields and started putting inputs exactly where the soil actually needed them, which turned out to be better for margins and the land.

Here is what should keep agricultural operators who wait to implement AI up at night: While you are deciding whether precision agriculture fits your operation, the farm across the county just cut their input costs by 30% and their water usage by 25%. Their yields are more consistent. Their livestock losses dropped to numbers that seemed impossible two years ago. They are landing the premium contracts you used to get.

The dairy that implemented AI herd monitoring six months ago catches mastitis before it spreads to the bulk tank. The orchard that added predictive pest modeling sprays less and loses less fruit. The grain operation that deployed yield mapping knows exactly which acres are profitable and which ones drag down the average. **They are not abandoning traditional farming. They are operators who decided that "we have always done it this way" was no longer acceptable when margins keep getting tighter and buyers keep demanding more.**

Your buyers have options. The industrial operations keep getting more efficient. The imported products keep getting cheaper. The premium markets keep raising their standards for consistency and traceability. **The relationships you have built over years with buyers, co-ops, and processors matter, but relationships do not survive indefinitely**

when quality consistency and documentation fall behind what they can get elsewhere. The buyer who has purchased from your family for decades will still take a meeting with the operation that can guarantee the specs they need, every delivery, with data to prove it.

The weather is not getting more predictable. The input costs are not getting cheaper. The labor is not getting easier to find. The environmental regulations are not getting simpler. **The only question is whether you will use AI to become the operation that combines generational land knowledge with the precision that modern markets demand, or whether you will keep telling yourself that experience alone will protect you from competitors who offer experience plus everything else.**

Your understanding of your land is not being automated. Your animal husbandry instincts are not being outsourced. Your ability to read the sky, walk the fields, and know in your gut when something is not right is not something a sensor network can replicate. But the operations that wrap that knowledge in AI-enhanced monitoring will produce more consistently. They will reduce their losses. They will build the reputations that command premium prices while everyone else sells into commodity markets.

Here is the truth that successful agricultural operations understand: **AI does not farm your land. It helps you see what you could not see before.** Every soil sensor reading reveals variation that walking the field would miss. Every livestock monitor catches the animal acting differently before you would notice on your morning rounds. Every yield map shows you which management decisions actually worked and which ones you have been doing out of habit. The technology extends your senses across acres you cannot physically cover, giving you information to make better decisions with the judgment only you can provide.

The operators who understand this are not abandoning their farming heritage. **They are becoming the kind of farmers their grandparents would have been if they had access to these tools, combining hard-won land knowledge with technology that makes every decision more informed.**

Your land, your livestock, and your crops do not care about technology trends. **They respond to good decisions made at the right time. AI just helps you make more of those decisions, faster, with information you could not gather on your own.**

How long before your buyers start noticing that the operation down the road delivers more consistent quality with documentation you cannot match?

Chapter 15

Construction & Trades

Technology That Supports Craftsmanship

Construction **contractors, electricians, plumbers, HVAC technicians, carpenters, and other skilled trade professionals** represent the most skeptical audience for AI implementation. Success depends entirely on proven skills and the ability to solve complex problems that no computer has experienced. These professionals have built careers based on knowledge gained from years of firsthand experience dealing with real-world challenges that textbooks cannot teach.

Yet, **customer expectations have shifted toward accurate project timelines, transparent pricing, and proactive communication,** which manual estimating and scheduling systems cannot consistently deliver. Competition from larger contractors with AI-powered project management, automated estimating, and predictive scheduling wins contracts through superior predictability, convenience, and reliability. The challenge is to leverage technology to meet modern project requirements while preserving the artistry and safety expertise that command premium pricing.

Success Story
BQ Electric

Brian Quinn founded *Best Quality Electric* (BQ Electric) in Hicksville, Long Island, in 2013. He **built his reputation on quality workmanship and reliability**, earning the trust of residential and commercial customers across Nassau and Suffolk Counties. His team of electricians took pride in solving complex wiring challenges. By 2018, the company had grown to the point where they finally hit $1.5 million in revenue.

Their business systems were simply described as "interesting." Take dispatching. When a technician left for a job, they would text the office the letter "D" for departing, followed by the customer's name. Someone in the office would then write that information on a whiteboard. "We were so high-tech," the operations manager jokes.

The whiteboard worked when they had three trucks. But Quinn knew that **growth would require something more sophisticated.** Customer expectations had shifted, people wanted accurate arrival times, transparent pricing, and proactive communication. Larger competitors with professional software were winning jobs by appearing more organized and reliable. And keeping track of inventory, job costs, and marketing results through spreadsheets and memory was becoming impossible.

The Question: How do you **modernize your operations** without losing the craftsmanship and personal service that earned your reputation in the first place?

They started researching field service management software. That is when they found ***ServiceTitan***, a platform built specifically for trades contractors like electricians, plumbers, and HVAC technicians. The results exceeded anything they had imagined.

After one year with *ServiceTitan*:

- **Revenue nearly doubled** from $1.5 million to $2.8 million.
- **Fleet expanded** from 3 trucks to 9.
- **Google reviews quadrupled** through automated post-service requests.
- Dispatching went from whiteboard scribbles to **real-time digital tracking.**

Within four years:

- Revenue grew to $5 million (233% growth from their 2018 baseline).
- Purchased a second commercial building.

- **Built a dedicated call center**, new offices, and conference rooms.
- **Created a training center** to develop the next generation of electricians.
- Set a new goal: $10 million within three to four years.

"Until we started ***ServiceTitan***, we didn't know that there was money on the table."

What changed?:

Efficient dispatching and customer updates. Appointments are shown on a virtual board, and technicians simply tap their tablets to check in. **ServiceTitan** then automatically sends customers an arrival time, setting BQ Electric apart from competitors who still make clients wait without updates.

Job tracking and billing. Technicians can review customer history, track materials used, and collect payment in the field using their tablets.

Marketing accountability. Through **ServiceTitan's Marketing Pro**, the company generates unique phone numbers for each advertising campaign. For the first time, they can see exactly which ads bring in revenue and which waste money. They also set up automated review requests—every completed job triggers a prompt asking satisfied customers to leave feedback.

Financial visibility. Now they generate detailed financial reports in moments, tracking everything from technician productivity to job profitability to marketing ROI.

Fleet management. With GPS tracking and dashboard cameras on every truck, they know where technicians are and can investigate any incidents. The visibility helps with scheduling, accountability, and even insurance.

The technology, averaging $200 per month per technician, handles the administrative complexity. The electricians handle the electrical work. Brian Quinn still makes the calls on tricky diagnostic challenges. His team still takes pride in workmanship that keeps customers coming back—BQ Electric has earned over 400 five-star reviews and has been named "Best Electrician on Long Island" for seven consecutive years.

Trust: Getting Buy-In from Skilled Tradespeople

The **successful adoption of AI in construction depends on its capacity to support, not threaten, the expertise and safety knowledge of skilled tradespeople.** It is essential to present AI as a sophisticated tool that empowers professionals to be more effective and efficient, but it is not a substitute for their roles.

Scripts for Craftsmanship Enhancement

For the primary electrician concerned about skill devaluation: "This project management AI tracks code requirements and material specifications so you can focus on the complex problem-solving and safety decisions that showcase your 20 years of experience."

For the experienced carpenter worried about job changes: "The scheduling and estimating tools handle the paperwork so you spend more time on actual construction and training apprentices in the skills that matter."

For the safety manager concerned about compliance: "AI monitoring helps catch potential safety issues before they become problems, supporting your expertise with comprehensive tracking that keeps everyone protected."

Ready: Preparing for Enhanced Project Management

Measure twice, cut once. The same principle applies to AI implementation, except **most contractors skip the measuring entirely.** They see a demo of some project management software, get excited about automated scheduling and instant estimates, and sign up without ever checking whether their business can actually use it. Three months later, the software sits untouched because nobody has time to enter data between jobs, and **the monthly subscription becomes an expensive reminder of good intentions.**

Here is where to begin to measure. Start with your estimating process. How do you build quotes right now? If the answer involves pulling numbers from your head based on twenty years of experience, you have a problem. Not because your estimates are wrong, they are probably pretty accurate, but because **AI estimating tools learn from historical job data.** They need to see what you quoted, what you actually spent -on materials and labor, and how long the job took compared to your original timeline. If that information lives in a filing cabinet full of handwritten invoices or scattered across three different spreadsheets that nobody updates, the AI has nothing to learn from. **Your first preparation task is not buying software; it is documenting your last fifty jobs in a format a computer can read.**

Next, look at your scheduling reality. Most contractors we talk to swear they have a scheduling system. Then, when we ask to see next week's job assignments, they pull out a whiteboard or open a text thread with their crew. **That is not a system; that is organized chaos that works only because you are personally holding it together.** AI scheduling needs structured data: job locations, estimated durations, crew certifications, equipment requirements, and customer availability timeframes. If you cannot answer those questions for every job on your calendar right now, you need to build that discipline before any software can help you optimize it.

Then get honest about your crew's technology comfort. Construction sites are tough on devices, cell coverage is unreliable, and your best electrician might be 65 years old and perfectly happy with the paper timesheets he has used for three decades. **None of that disqualifies you from AI implementation, but it shapes what you should implement first.** A crew that does not use a mobile app will not benefit from real-time job tracking, no matter how good the software is. **Start with tools that solve problems your team actually cares about, like eliminating the end-of-day paperwork they hate, then build from there.**

Safety documentation deserves its own hard look. Regulations keep getting more complex, and the contractors who think they are compliant often discover gaps when someone actually audits their records. AI can help with safety tracking and compliance monitoring, but only if you have a baseline. When was your last toolbox talk documented? Are your crew certifications current and recorded in an accessible location? Do you have incident reports from the past two years in a searchable format? **If the answer to any of these is "probably somewhere," that is your starting point.** AI safety tools amplify good documentation habits; they do not create them from nothing.

Finally, **talk to your accountant before you talk to vendors.** Construction businesses have complex job costing, and AI project management tools need to integrate with how you track profitability. If your books are a mess, if you do not actually know which jobs made money and which ones bled you dry, no amount of AI optimization will fix the underlying problem. **Get your financial house in order first.** You need to understand your real costs per job type, your actual labor burden rates, and your true overhead allocation before any AI tool can help you bid smarter.

The contractors who win with AI are the ones who do the unglamorous preparation work first. They measure twice. Then they cut once, and the cut is clean.

Advance: Putting AI to Work

Integrating advanced tools like ***ServiceTitan AI*** into daily operations highlights the **tangible benefits of adopting technology-driven strategies**—from streamlining workflows and enhancing team coordination to proactively addressing safety and maintenance needs. This methodical progression not only supports consistent craftsmanship but also sets the stage for meaningful efficiency gains and superior client engagement throughout each project phase.

Three Best AI Programs for Construction and Trades

1. ServiceTitan for Field Service Trades - The Service Call Excellence Engine

What it does: AI-powered job scheduling and dispatch, customer history tracking, inventory management, technician performance analytics, and automated follow-up communication.

Why service-oriented trades choose it: ServiceTitan combines field service management with business intelligence designed specifically for plumbing, HVAC, and electrical service companies.

Real impact: Service contractors report a 67% improvement in first-call resolution and an 89% increase in upselling success from better preparation and exposure to customer insights.

Monthly cost: $125-398+ per month, depending on company size and features. Small operations (1-5 techs) start around $125/month, while larger operations (20+ technicians) typically pay $250-500+ per month/per technician.

Best for: Service-oriented trades focused on residential and commercial maintenance, repair, and installation work.

2. *Procore with AI Project Intelligence* - The Complete Construction Management Brain

What it does: AI-powered project scheduling, cost forecasting, safety compliance tracking, quality control monitoring, and comprehensive project communication.

Why construction professionals choose it: Procore is designed specifically for construction with industry-specific workflows for project management, safety documentation, and team coordination.

Real impact: Construction companies report 25% reduction in project delays and 34% improvement in budget accuracy through AI-enhanced planning and tracking.

Monthly cost: Custom pricing based on annual construction volume, typically 0.1-0.2% of project hard costs. Small contractors can expect $4,500-10,000 annually, mid-size firms $10,000-25,000, and large operations $25,000-60,000+.

Best for: Medium to large construction companies managing multiple complex projects simultaneously.

3. *Jobber with AI Scheduling* – The Home Service Optimizer

What it does: Smart job scheduling, automated customer communication, inventory tracking, quote generation, and team coordination designed for residential service businesses.

Why home service professionals adopt it: Jobber provides comprehensive management specifically designed for service contractors with easy adaptation and powerful AI features.

Real impact: Service businesses using Jobber report 45% reduction in scheduling conflicts and 67% improvement in customer communication through automated updates and proactive reminders.

Monthly cost: $39-599 per month depending on team size and features. Solo operators start at $39/month, while team plans range from $169-599/month for 5-15 users. Additional users cost $29/month each. Plus payment processing fees of 2.9% + $0.30 per transaction.

Best for: Small to medium-sized businesses focused on residential customers and building strong relationships.

Connect: Making AI and Trade Expertise Work Together

The turning point for construction comes when AI analytical power combines with human trade expertise to create project capabilities that neither could achieve alone.

The Complete Construction Integration Pattern

The best results come from construction companies that adhere to the same integration pattern:

AI Handles:

- Project scheduling
- Material tracking
- cost estimation
- Safety compliance monitoring
- Customer communication
- Performance analytics

Human Expertise Provides:

- Complex problem-solving
- Craftsmanship quality
- Safety judgment
- Customer consultation
- Application of trade-specific knowledge

Working Together Results:

- Teams experience
- More accurate project planning
- Enhanced safety compliance
- Reliable scheduling
- Preserved craftsmanship standards

Real-World Integration Success

Project Planning Enhancement:

AI Analysis: System analyzed historical project data and current material costs **predict accurate timelines and budgets** for commercial electrical installation.

Trade Expertise: Expert electrician Mike applied his 18 years of experience to review AI estimates and adjusted for site-specific challenges and local code requirements.

Superior Outcome: Experienced 15% more accurate estimates with 34% faster completion through optimized material ordering and crew scheduling.

Safety Compliance Integration:

AI Monitoring: Ongoing supervision of safety protocols, equipment inspections, and compliance with regulatory standards across multiple job sites.

Safety Expertise: An **experienced safety manager interpreted AI data** within the context of specific site conditions and weather factors to make adjustments directly affecting safety.

Result: No safety incidents, 45% less compliance documentation time, and better regulatory audit results.

Building Customer and Team Confidence

Team Response:

- 91% of **electricians expressed confidence in AI-enhanced project management.**
- Senior **staff recognized AI as a tool for displaying their proficiency** more effectively.

- **Apprentices appreciated structured learning** opportunities with AI-assisted training.
- Administrative **staff welcomed a reduction in job estimating and scheduling** coordination.

Customer Response:

- An 88% **positive reaction to enhanced project communication** and timeline accuracy.
- **Clients appreciated transparency** about project progress and potential delays.
- **No concerns were raised about work quality or safety standards.**
- Several **customers requested information about predictive maintenance services.**

Keep: Maintaining Construction Excellence

Construction **businesses that successfully implement AI create advantages that strengthen over time.** They do this by establishing systems that enhance project management while preserving the craftsmanship reputation and customer relationships that define their value.

AI implementation creates opportunities for crew advancement that strengthen your competitive position:

- **Technology leadership roles**: Your most adaptable crew members become AI system champions who train others and troubleshoot problems.
- **Career progression paths**: Employees who master AI tools become more valuable, creating retention incentives beyond hourly wages.
- **Apprenticeship enhancement**: New workers learn both traditional trade skills and modern technology, accelerating their path to productivity.
- **Safety expertise development**: AI monitoring creates data that experienced workers can interpret, combining technology with trade knowledge.

Customer Communication Excellence

AI-enhanced communication becomes a competitive differentiator when maintained consistently:

- **Project updates**: Automated progress reports keep customers informed without requiring your time during busy job phases.
- **Schedule transparency**: Real-time scheduling visibility reduces the "when will you be here" calls that interrupt productive work.

Documentation quality: Comprehensive project records protect against disputes and demonstrate professionalism that justifies premium pricing.

Referral generation: Customers who experience superior communication become advocates who recommend you to neighbors and colleagues.

Market Position Through Innovation

- **Frame AI adoption as a commitment to quality** and customer service, not just efficiency gains.
- **Use technology capabilities** to pursue larger, more complex projects while maintaining service standards.
- Demonstrate how **AI improves safety records** and project outcomes, reinforcing your reputation for reliability.
- Build recognition as a **contractor who combines traditional craftsmanship with modern project management.**

The Construction and Trades Advantage

The most successful contractors do not treat AI as a replacement for the craftsmanship that built their reputation or a shortcut around the hard-won expertise their crews bring to every job. They **use it as an operations amplifier that requires experienced tradespeople to fully realize its potential.**

Key Success Elements:

- AI handles **estimating calculations, scheduling optimization, safety documentation, and customer communications.**
- Crews provide skilled craftsmanship, on-site problem-solving, quality judgment, and the customer relationships that generate referrals.
- Customer r**elationships remain personal while project capabilities expand to compete for larger jobs.**
- Safety stays tight because **AI handles the documentation that used to get skipped** during busy seasons.

Construction excellence is not about choosing between proven trade methods and modern technology. **It is about using AI to make experienced contractors more accurate, more reliable, and more profitable than ever before.**

Contractors thoughtfully using AI create advantages that large commercial operations and fly-by-night competitors cannot replicate. **The combination of precise estimating, reliable scheduling, and genuine craftsmanship takes years to develop.** A low-ball bid is just a number. A contractor who shows up when promised, communicates throughout the project, catches problems before they become expensive, and delivers work that lasts for decades? That is trade expertise no software can fake.

The electrical contractor who implemented AI estimating did not lose his ability to read a job site. **He stopped leaving money on the table with manual takeoffs and started winning bids at margins that actually made sense.** The plumbing company that added scheduling optimization did not become impersonal. They stopped double-booking crews and started showing up when they said they would, which turned out to be the competitive advantage their customers cared about most. The general contractor who automated project documentation did not sacrifice quality. He stopped losing disputes over change orders and started building the paper trail that protects everyone when problems arise.

Here is what should keep contractors who wait to implement AI up at night: While you are debating whether technology fits your boots-on-the-ground business, the contractor across town just delivered an accurate estimate in two hours. Yours took two days. By the time you sent your bid, they had already scheduled the job.

The HVAC company that implemented AI scheduling six months ago never misses appointment windows. The roofing contractor that added safety monitoring has an incident rate that makes insurance companies smile. The remodeler who automated customer updates gets five-star reviews mentioning communication, not just craftsmanship. **They are not becoming corporate outfits. They are contractors who decided that "I will get back to you" was no longer acceptable when customers can get instant quotes from three competitors on their phone.**

Your customers have options. The franchise operations keep expanding. The tech-enabled startups keep improving their platforms. The big commercial contractors keep moving into residential work. **The relationships you have built over years matter, but relationships do not survive indefinitely when scheduling reliability and communication fall behind what customers can get elsewhere.** The homeowner who has trusted you for a decade will still get three bids on the next project if your estimate takes a week and theirs arrived in an hour.

The labor shortage is not getting better. The material costs are not getting more predictable. The customer expectations are not reverting to what they were before smartphones made instant communication standard. **The only question is whether you will use AI to become the contractor who combines genuine craftsmanship with the accuracy and responsiveness that modern customers demand, or whether you will keep telling yourself that quality work alone will protect you from competitors who offer quality work plus everything else.**

Your trade skills are not being automated. Your problem-solving instincts are not being outsourced. Your ability to look at a job site and see what needs to happen is not something software can replicate. But the contractors who wrap that expertise in AI-enhanced operations will win more bids. They will complete more projects on time. They will build the reputations that command premium prices while everyone else races to the bottom on price.

Here is the truth that successful contractors understand: **AI does not replace the work. It protects the work.** Every accurate estimate prevents a job that loses money. Every optimized schedule prevents a crew from standing idle. Every safety document prevents a violation that shuts down a site. Every customer update prevents the phone call that interrupts your best installer in the middle of a critical task. The technology handles what used to fall through the cracks, so your team can focus on what they actually do best.

The contractors who understand this are not abandoning their trade values. **They are building the kind of companies they always wanted to run, with AI handling the chaos so they can focus on the craftsmanship that made them pick up tools in the first place.**

How long before your best customers start noticing that the other contractor always shows up on time and never surprises them with the final bill?

Chapter 16

Service Industries
AI That Scales Personal Touch

Specialized service businesses face a unique challenge. **You built your reputation on personal relationships, hands-on expertise, and being embedded in your local area.** Whether you run an **auto repair shop** where customers trust you with their family's safety, a **dance studio** where you shape children's confidence, a **pet boarding** facility where animals become part of your extended family, or a **marina** where you are part of a tight-knit boating community, your success depends on trust that cannot be automated.

Success Story
Top Shop Auto

When *Top Shop Auto* in Lafayette, California implemented ***Shopmonkey*** in their auto repair shop, owner Frank Powell faced immediate pushback from mechanics who had spent years mastering their craft. They worried that automation meant their expertise no longer mattered.

Frank's response: "This software does not fix cars—you do. But it does eliminate the running back and forth to the office, the lost paperwork, the time you waste tracking down parts orders instead of turning wrenches. We are not replacing mechanics with computers. We are giving you **tools that let you do more of what you are good at**."

He was right. Within two weeks, *Top Shop Auto* saw 25% efficiency gains—not from mechanics working faster, but from eliminating the administrative tasks that pulled them away from actual repair work. **Mechanics spent their time diagnosing and fixing vehicles instead of hunting for work orders or calling customers for approval.**

Revenue grew 20% without adding staff. Employee satisfaction improved because technicians were doing more skilled work and less paperwork. Frank proved his point: AI enhanced their expertise rather than replacing it.

Trust: Getting Buy-In from Service Professionals and Customers

Service industries add unique concerns to trust. Workers worry that automation will devalue the relationships they have built with regular customers, business owners fear losing the personal reputation that generates referrals, and customers question whether AI can deliver the individualized attention and flexibility that justify premium pricing.

The successful adoption of AI in service businesses depends on **presenting technology as a tool** that enhances rather than replaces personal relationships and professional expertise. You must show employees and customers that AI handles coordination and communication so service professionals can focus on delivering exceptional personal attention.

Scripts for The Service Industry

For the experienced service professional worried about customer relationships: "This scheduling AI handles booking confirmations and appointment reminders so you can spend more time learning customer preferences and delivering the personalized service that keeps them loyal."

For the team member concerned about job security: "The AI system manages coordination and paperwork, so you focus on what you do best—building relationships with customers who request you specifically because they value your expertise."

For the quality-focused employee worried about consistency: "AI tracking helps ensure every customer receives the same attention to detail and personalized service, supporting your commitment to excellence with comprehensive preference tracking."

Scripts for Building Customer Trust

For regular customers concerned about losing personal service: "AI helps us remember your preferences, schedule history, and special requests so you receive even more personalized attention from professionals who know exactly what you need."

For new customers worried about quality: "Our AI scheduling ensures the right professional for your specific needs, and our quality tracking system maintains the high standards that have earned us hundreds of five-star reviews."

For price-sensitive customers questioning value: "Better scheduling efficiency and reduced coordination time help us provide more reliable service and faster response times without raising prices."

Building Employee Confidence

Employees do not fear technology itself, but the risk of being replaced by it. This distinction is relevant to all services.

For many workers, technology offers opportunities for growth and greater efficiency, but concerns arise when automation or new systems threaten their job security. The apprehension is not about adopting advanced tools, but rather about the uncertainty regarding future employment and changes to traditional roles.

Success Story
Synergy Dance Academy

At *Synergy Dance Academy*, owner Kari Fisher faced skeptical instructors who worried that online systems would make the studio feel impersonal. Her message was clear: "Parents want easy registration and communication, but they want their children taught by people who care about their development. ***Jackrabbit Dance*** handles the administrative work so you can spend more time actually teaching."

The result: instructors spent less time on billing questions and scheduling conflicts, more time coaching students. Parent satisfaction increased because communication improved, not because technology replaced personal interaction.

Addressing Customer Concerns

Your customers do not wake up wishing their service provider would adopt more technology. They **want the same reliable service they have always received, just easier to access and coordinate.**

Synergy Dance manages 15 routines and 15 costumes per student in a growing studio. After implementing AI, parents repeatedly told the owner, "That was so easy." The platform handles online registration at any time of day, automated billing and payment collection, costume and equipment management across hundreds of items, and a parent communication portal that shows class schedules, attendance records, and skills progression. Instructors spend less time on billing questions and more time teaching.

Ready: Preparing Your Operation

You have **tight margins, small teams, and customers who will notice immediately if service quality drop**s during transition. This means preparation for AI implementation is not optional; it is critical.

Service businesses have the messiest technology situations of any industry we work with. Your customer data lives in three different places: the scheduling software you bought five years ago, the spreadsheet your office manager maintains, and the head of your longest-tenured employee. No AI system is going to extract that information automatically.

Before you buy anything, **do an honest inventory of where your business information actually lives.** Pull out your customer list. Is it current? Does it include contact preferences, service history, and the details that make your service personal? **If that information exists only in conversation and memory, your first job is not buying AI.** Your first job is documenting what you already know so that AI has something to work with.

One campground owner in Colorado learned this the hard way. She implemented ***Campspot*** for online reservations, expecting it to handle everything. What she discovered was her repeat campers had preferences. The family that always wanted site 14 near the bathhouse, the RV group that needed pull-through sites close together, the solo hiker who preferred the quiet loop away from generators. **None of that was written down anywhere.** Her staff spent the first month fielding complaint calls from loyal customers who got assigned random sites by the new system. She ended up building a customer preference database from scratch before the AI could actually help. Six weeks of extra work that could have been done before implementation.

Talk to your front-line staff before you talk to AI vendors. They know exactly where time gets wasted, and when customers get frustrated. They also know which parts of your current system actually work well and should not be touched.

Be honest about your team's comfort with technology. Service businesses often have staff spanning three generations. Your AI implementation will fail if you choose tools that half your team cannot or will not use. **The fanciest system in the world is worthless if your best employee refuses to check it.**

Finally, identify your single biggest operational headache. Not your three biggest headaches, **just your one biggest headache.** Is it phone tag with customers? No-shows that leave gaps in your schedule? How many hours spent on invoicing? **Pick the problem that costs you the most time, money, or customer goodwill.** That is where you start. The businesses that succeed with AI solve one real problem, prove the value, and expand from there.

Your customers chose you because you know them personally. **AI preparation is about making sure technology supports those relationships instead of getting in the way.**

Advance: Putting AI to Work Across Service Industries

Success Story *Marsh Cleaning*

*Cynthia** had operated her residential cleaning company in Portland, Oregon for 14 years, building a team of twenty-three cleaning professionals who served 340 regular customers across the metropolitan area. **Her business grew through word-of-mouth referrals based on consistent quality, trusted relationships,** and the kind of personalized service that made customers comfortable giving her team keys to their homes.

The service landscape was evolving rapidly:

- Customer expectations had shifted toward **instant online booking, accurate arrival times, and proactive communication** that manual scheduling systems could not deliver consistently.
- Competition from **national cleaning franchises with AI-powered scheduling, automated reminders, and standardized quality control was winning contracts** through superior convenience and reliability.
- **Staff coordination was becoming increasingly complex** as customer preferences, special requests, and schedule changes overwhelmed manual tracking systems.

- **Quality consistency challenges were intensifying** as the business grew, making it difficult to maintain the personalized service standards that justified premium pricing and generated referrals.

The dilemma: How do you leverage technology to meet modern customer expectations and scale your business while preserving the personal relationships and service quality that built your reputation?

Three Best AI Programs for Service Industries

1. *Jobber with AI Scheduling and Route Optimization* - The Complete Home Service Business Brain

What it does: Smart job scheduling, automated customer communication, inventory tracking, quote generation, and team coordination designed for residential service businesses.

Why service professionals adopt it: Jobber provides comprehensive business management specifically designed for home services, cleaning, landscaping, and property maintenance with intuitive interfaces and powerful automation.

Real impact: Service businesses using **Jobber** report a 52% reduction in scheduling conflicts and a 67% improvement in customer communication, and a 34% increase in job completion rates through better coordination.

Monthly cost: $39-599 per month depending on team size and features. Solo operators start at $39/month, while team plans range from $169-599/month for 5 -15 users. Additional users cost $29/month each. Plus payment processing fees of 2.9% + $0.30 per transaction.

Best suited for: Home services, cleaning companies, landscaping businesses, and property maintenance providers focused on residential customers.

2. *Vagaro with AI Booking* - The Personal Services Business Brain

What it does: Intelligent appointment scheduling, automated booking and reminders, customer preference tracking, marketing automation, payment processing, and comprehensive client management.

Why service professionals choose it: **Vagaro** is designed specifically for salons, spas, personal trainers, pet groomers, and wellness professionals with features that enhance customer relationships and service delivery.

Real impact: Personal service businesses report a 73% reduction in no-shows, a 45% increase in repeat bookings, and an 89% improvement in client satisfaction through automated reminders and personalized service tracking.

Monthly cost: $25-250, depending on staff size and features.

Best for: Salons, spas, personal trainers, massage therapists, pet groomers, and wellness professionals focused on building long-term client relationships.

3. *Housecall Pro with Smart Dispatch* – The Field Service Excellence Engine

What it does: AI-powered scheduling and dispatch, customer history tracking, automated communication, job tracking, payment processing, and comprehensive business analytics for mobile service businesses.

Why repair and maintenance professionals choose it: Housecall Pro combines field service management with business intelligence designed for businesses that travel to customers for repairs, installations, and maintenance work.

Real impact: repair and maintenance professionals choose it: Housecall Pro combines field service management with business intelligence designed for businesses that travel to customers for repairs, installations, and maintenance work.

Monthly cost: $49-199, depending on team size and features.

Best for: Appliance repair, auto repair, handyman services, pest control, pool services, and other mobile service businesses that require efficient dispatch and route optimization.

Automotive Services: Diagnostic AI and Customer Communication

Auto repair, body shops, tire centers, and detailing services share a fundamental challenge: customers do not understand what you do. They cannot see the work, do not know if the diagnosis is accurate, and fear being overcharged.

This creates friction that slows work approvals, reduces upselling, and can damage customer relationships

Three Best AI Programs for Automotive Services

1. *Shopmonkey* handles the complete customer journey from initial contact through final payment.

What it Does: Technicians take photos of problems, annotate them with explanations, and text them directly to customers. Instead of describing brake wear over the phone, mechanics send photos showing exactly what 30% looks like and why it matters now.

Why auto shops choose it: Technicians take photos of problems, annotate them with explanations, and text them directly to customers. Instead of describing brake wear over the phone, mechanics send photos showing exactly what 30 percent looks like and why it matters now. *Top Shop Auto* saw customer approval rates jump to 75 percent using this feature—far higher than phone-based approval. *Windows & Wheels Auto Detailing* reduced no-shows by 90 percent using text and email reminders.

Real impact: *Top Shop Auto* saw 25% efficiency gains and 20% revenue growth without adding mechanics.

Monthly cost: Pricing starts around $199/month for small shops. Payment processing and advanced features add cost, but shops report the software pays for itself within the first month through increased efficiency and reduced no-shows.

Best for: Independent auto repair shops, collision centers, tire shops, and auto detailing businesses with 2-20 employees looking to modernize customer communication, reduce phone time, and increase technician productivity.

2. *Tekmetric Shop Management* – the Shop Data Insight Pro

What it does: ***Tekmetric*** provides comprehensive shop management with digital vehicle inspections, real-time reporting, automated workflows, and streamlined estimate and authorization processes.

Why auto shops choose it: ***Tekmetric*** offers powerful real-time data insights that help shop owners track technician hours, employee productivity, and profitability. The platform integrates with SEO tools to track marketing ROI and fine-tune campaigns. Shops get comprehensive financial tracking, automated payment collection, and the ability to identify areas for improvement through detailed performance reports.

Real impact: *Mission Auto KC* doubled monthly revenue from $40,000-$50,000 to $90,000 and achieved 20% higher gross profit. ***Tekmetric*** users see a 25% average increase in repair orders within the first 120 days.

Monthly cost: Pricing is $199-$250/month for small shops with additional fees for payment processing and advanced features.

Best for: Auto repair shops that want powerful real-time data insights, comprehensive reporting capabilities, and the ability to track technician productivity, financial metrics, and marketing ROI all in one platform.

3. *Mitchell 1* AI Diagnostic - The Auto Shop Master Investigator & Estimator

What it does: ***Mitchell 1*** provides diagnostic software and repair information systems with comprehensive vehicle repair data, labor time guides, and technical service bulletins to help technicians diagnose and repair vehicles accurately.

Why auto shops choose it: When technicians encounter unfamiliar vehicles or complex repairs, ***Mitchell 1*** provides the technical information they need to diagnose problems confidently and estimate repairs accurately. The system includes wiring diagrams, component locations, diagnostic trouble code information, and step-by-step repair procedures.

Real impact: Auto repair shops using ***Mitchell 1*** report faster diagnostic times, more accurate estimates, and improved technician confidence when working on unfamiliar vehicles or complex repairs. The system reduces diagnostic errors and helps shops provide more accurate quotes to customers.

Monthly cost: Pricing starts around $295/month for basic diagnostic software access, with additional costs for enhanced features and integrations.

Best for: Auto repair shops that need comprehensive repair information, diagnostic support, and technical resources to handle a wide range of vehicle makes and models confidently.

Personal Care & Grooming Services

For **hair salons, nail salons, barber shops, spas, and pet grooming businesse**s, common appointment challenges can seriously disrupt daily operations and customer satisfaction. Double-booking often leaves customers feeling neglected, which can lead to frustration and lost trust. No-shows waste valuable time slots and reduce income, as staff are left with gaps in their schedules. Administrative chaos, such as missed appointments or billing errors, damages client relationships and can make your business appear unprofessional. By implementing efficient scheduling systems and sending automated reminders, you can minimize these issues and ensure smooth operations.

Success Story
Salon 77 West

Salon 77 West in Pleasanton, California, used ***Vagaro*** and generated $4,000 in extra revenue from one Daily Deal campaign while securing 38 new clients. The salon achieved a 37% revenue increase from the previous month. Across the industry, salons using ***Vagaro's*** Pay Later feature see a 30% increase in checkout conversion, 41% increase in average order value, and 45% higher purchase frequency.

Three Best AI Programs for Salons, Spas & Beauty Services

1. *Vagaro Salon Management* – The Premier Salon Operations Leader

What it does: ***Vagaro*** serves hair salons, nail salons, spas, and pet grooming businesses with online booking, automated appointment reminders, payment processing, client management, and marketing automation all in one platform.

Why professionals choose it: Clients book appointments 24/7 through the business website, social media pages, or the ***Vagaro*** marketplace. The system sends automated reminders via text and email, dramatically reducing no-shows. ***Vagaro*** handles recurring appointments, packages, memberships, and retail sales seamlessly. The Pay Later feature increases checkout conversion by 30% by letting clients book now and pay later.

Real impact: *Salon 77 West* generated $4,000 in extra revenue from one campaign with a 37% revenue increase. Salons using online booking typically see a 30% increase in sales, with over 60% of online bookings made outside regular working hours. The platform allows businesses to stay "open" 24/7 even when doors are closed.

Monthly cost: ***Vagaro*** charges a monthly subscription starting around $25/month plus transaction fees for payment processing, making it one of the most affordable options for small service businesses.

Best for: Hair salons, nail salons, barber shops, spas, and pet grooming businesses that want an all-in-one platform for booking, payments, client management, and marketing at an affordable price point.

2. *Fresha* – The Salon & Spa Appointment Booking Guru

What it does: ***Fresha*** provides subscription-free business management software for salons and spas, offering online booking, appointment management, point-of-sale, client records, and marketing tools without monthly software fees.

Why professionals choose it: ***Fresha*** charges no monthly subscription fees—businesses pay only for transaction processing and optional marketing features. The platform serves over 190,000 businesses globally and provides professional management tools without the traditional software costs. ***Fresha's*** online booking system optimizes salon schedules with real-time availability, and over 80% of appointments at many salons are booked online directly by clients.

Real impact: Salons using ***Fresha*** report an average 30% increase in revenues through online booking. The platform's marketplace connects salons with new clients, and 74% of bookings through the marketplace are first-time visitors. Salons save money by eliminating monthly software costs while still getting full-featured management tools.

Monthly cost: No monthly subscription fee. ***Fresha*** generates revenue through payment processing fees and optional premium features, making it accessible for startups and small businesses.

Best for: Salons, spas, and wellness businesses that want full-featured management software without monthly subscription costs, especially those just starting out or operating on tight margins.

3. *GlossGenius Business Management* – The Soloprenuer Master Marketer

What it does: ***GlossGenius*** provides business management software specifically designed for independent beauty professionals, offering scheduling, payments, client management, and marketing automation tailored to solo practitioners and small teams.

Why professionals choose it: ***GlossGenius*** focuses on the specific needs of independent beauty professionals who operate as solo entrepreneurs or small teams. The platform handles everything from Instagram booking integration to automated review requests, making it easy for independent professionals to look and operate like established businesses. The interface is designed for professionals who work from home studios, rent chairs, or run small independent salons.

Real impact: Independent professionals using ***GlossGenius*** report better client retention through automated reminders and follow-ups, increased booking rates through social media integration, and more professional client communications that help them compete against larger salons.

Monthly cost: Pricing starts around $24/month for solo professionals with transaction fees for payment processing.

Best for: Independent hairstylists, nail technicians, estheticians, and other solo beauty professionals who need business management tools designed specifically for independent practitioners, not multi-location salons.

Activity Centers: Dance, Martial Arts, Music, and Fitness Studios

Dance studios, martial arts schools, music academies, gymnastics centers, and **art studios** face operational chaos: constantly changing schedules, complex billing with discounts and package pricing, costume and equipment management, and parent communication that never stops. This administrative burden keeps owners and instructors trapped behind desks when they should be teaching.

Three Best AI Programs for Studios & Activity Centers

1. *Jackrabbit Dance* – The Activity Center Manager

What it does: *Jackrabbit Dance* serves as the operating system for dance studios, martial arts schools, and similar education businesses with online registration, automated billing, costume management, and parent communication portals.

Why activity centers choose it: Parents enroll students at their convenience, eliminating phone tag and office-hour constraints. The system handles recurring payments, pro-rating for mid-month starts, automatic late payment reminders, and flexible discount structures that eliminate billing confusion. For studios managing massive inventory—like *Synergy Dance Academy* with 15 costumes per student—***Jackrabbit Dance*** organizes all costume data in one system.

Real impact: *Synergy Dance Academy* parents love the ease of registration and access to attendance records and skills progression. Studios report reduced administrative time, fewer billing errors, and improved parent satisfaction through better communication and transparency.

Monthly cost: Pricing is based on enrolled students, typically starting around $89/month for small studios (under 75 students) and scaling up with enrollment.

Best for: Dance studios, martial arts schools, gymnastics centers, music schools, tutoring centers, and similar education businesses with recurring enrollments, complex scheduling, and significant parent communication needs.

2. *Pike13 AI* – The Class Management Expert

What it does: ***Pike13*** provides business management software for fitness studios, martial arts schools, and activity centers with scheduling, billing, client management, and mobile-friendly tools accessible anytime, anywhere.

Why activity centers choose it: ***Pike13*** was designed specifically for service-based businesses where instructors and staff need to manage operations on the go. The system provides real-time visibility into business performance with analytics and reporting tools.

Real impact: *Fit4MOM* franchise owners report that ***Pike13*** saves staff an average of 10 hours per week, significant for small teams. Franchise locations saw a drastic increase in follow-through from first-time class enrollments after switching to ***Pike13***. The company's 235 franchises and more than 1,400 class locations depend on ***Pike13*** for operational management.

Monthly cost: Pricing varies based on business size and features needed, typically competitive with ***Jackrabbit Dance***, with tiered plans for different business scales.

Best for: Fitness studios, martial arts schools, tutoring centers, and activity centers that need mobile-first management tools and real-time business insights, especially multi-location operations, or franchise businesses.

3. *Mindbody Business Management* – The Fitness Class Marketplace

What it does: *Mindbody* provides comprehensive business management and marketing software for fitness studios, yoga centers, wellness businesses, and activity centers with scheduling, client management, payment processing, and integrated marketplace exposure.

Why activity centers choose it: *Mindbody* operates one of the largest wellness marketplaces, connecting studios with potential clients actively searching for classes. The platform handles everything from class scheduling and attendance tracking to retail sales and staff management.

Real impact: Studios using ***Mindbody*** gain access to millions of potential clients through the marketplace while managing operations through comprehensive internal tools. The platform's widespread adoption means clients are already familiar with the booking interface, reducing friction for new students.

Monthly cost: Pricing varies significantly based on business size, features, and marketplace participation, typically starting around $129/month for small studios with additional fees for advanced features and marketplace exposure.

Best for: Yoga studios, Pilates centers, fitness studios, and wellness businesses that want both comprehensive management tools and exposure to a large marketplace of potential clients actively searching for classes.

Recreation & Outdoor Hospitality: Marinas, RV Parks, and Campgrounds

Recreation and entertainment businesses, including **marinas, RV parks,** and **campgrounds**, operate with complex inventory (slips, sites, hookups), seasonal demand fluctuations, and customers who expect both personal service and digital convenience. Peak season chaos and booking inefficiencies directly impact revenue potential.

Success Story
MacDougall's Cape Cod Marine Service

MacDougalls' Cape Cod Marine Service in Falmouth Harbor implemented ***Dockwa*** and saw a 525% increase in transient slip revenue between 2015 and 2018. The marina achieved better inventory management tracking, which allows seasonal slips to be available for transient guests when seasonal customers are out cruising. They "killed the phone" by requiring all transient reservations through ***Dockwa***, eliminating booking chaos during peak season—with excellent customer service through the platform.

Three Best AI Programs for Outdoor Hospitality

1. *Dockwa Marina Management* – The Boaters' Dream Guide

What it does: ***Dockwa*** serves as the operating system for marina management, handling reservations through contracts through payment with online slip reservations, digital contracts, visual marina maps, and automated payment processing.

Why marinas choose it: Boaters search availability, request reservations, and receive instant confirmation without calling the marina office which is critical during peak season when phone lines are overwhelmed. ***Dockwa*** allows marinas to bulk-send contracts to seasonal boaters, collect electronic signatures, and automatically process payments.

Real impact: *MacDougalls' Cape Cod Marine Service* saw 525% increase in transient slip revenue. *Edgartown Harbor* achieved 44% increase in annual nights served and revenue over three years, extended season by a full month, and 40% of all reservations booked on opening day because the system handled volume that would have overwhelmed phone-based booking.

Monthly cost: Pricing varies based on marina size and features. Basic reservation management starts around $300/month for smaller marinas, scaling up for larger operations or premium features.

Best for: Marinas and yacht clubs managing slip reservations, transient dockage, contracts, and payments.

2. *Campspot Campground Management* – The Overnight Outdoor Marketplace

What it does: ***Campspot*** provides comprehensive campground management software with online booking, dynamic pricing, site optimization, reservation management, and marketplace exposure for RV parks and campgrounds.

Why campgrounds choose it: ***Campspot*** includes dynamic pricing that adjusts rates based on demand with campgrounds using this feature earning 83% more revenue on average in 2024. The platform's grid optimization, site locking, custom booking rules, and heat map reporting help campgrounds maximize revenue from every available site night. The ***Campspot*** Marketplace connected campgrounds with over 106,000 new campers in 2024, with 74% being first-time visitors to those campgrounds.

Real impact: One campground achieved a 4x monthly revenue increase after implementation. Another saw a 50% increase in revenue during the summer months.

Monthly cost: Pricing varies based on campground size and features, with basic plans and scaling options for larger operations.

Best for: RV parks and campgrounds wanting comprehensive management tools with dynamic pricing capabilities and access to a large marketplace of campers actively searching for sites.

3. *RoverPass Reservation Management* – The RV Property Management Leader

What it does: ***RoverPass*** provides reservation management and marketing software specifically designed for RV parks and campgrounds with online booking, channel management, and property management tools.

Why campgrounds choose it: ***RoverPass*** focuses specifically on the needs of RV parks and campgrounds rather than trying to serve multiple hospitality segments. The platform handles everything from reservation management to housekeeping coordination. The system integrates with multiple online travel agencies and booking sites to maximize exposure.

Real impact: Campgrounds using ***RoverPass*** report improved reservation management, better utilization of available sites, and increased bookings through multi-channel distribution that puts their campground in front of travelers searching across different platforms.

Monthly cost: Pricing varies based on campground size and features, competitive with other campground management platforms.

Best for: RV parks and campgrounds that want campground-specific management tools with strong reservation management capabilities to distribute inventory across multiple booking platforms.

Connect: Making AI and Human Expertise Work Together

AI does not replace skilled employees in community and specialized services—it amplifies what they do best by eliminating what wastes their time. The **businesses succeeding with AI understand a fundamental truth: customers do not care about your technology.** They care about the person fixing their car, teaching their child, caring for their pet, or helping them dock their boat. AI's job is to make that person more effective, not more efficient at the expense of quality.

This means your **AI implementation needs clear guidelines** about what AI handles versus what requires human judgment and expertise.

AI Handles:

- Scheduling and confirmations
- Payment processing and billing
- Routine customer communications and reminders
- Inventory tracking and order alerts
- Basic information requests
- Administrative documentation

People Provide:

- Complex diagnostics and solutions
- Emotional support and rapport
- Service recommendation decisions
- Training and coaching
- Resolving complaints
- Community engagement

Working Together: Creates **faster response times** without sacrificing personal attention and gives your customers more direct contact with your team. Less time is spent on paperwork. Consistent communication strengthens client relationships. Your professional operation feels personal, not corporate, because of your ability to serve more customers without compromising quality.

Keep: Maintaining Service Excellence

AI implementation is not a one-time project; it is **an ongoing process of refinement.** Specialized service businesses that succeed long-term treat their AI systems like they treat their equipment: essential tools that require regular maintenance and improvement.

This ongoing process of improvement ensures that **AI remains effective in meeting business goals and adapting to changing market demands**. By prioritizing maintenance and development of AI tools, businesses can deliver better services, reduce operational risks, and stay competitive within their industries.

Quarterly System Reviews

Every quarter, evaluate three areas:

1. **What is working well that we should expand?**

At *MacDougalls' Cape Cod Marine Service*, the decision to "kill the phone" for transient reservations was controversial initially. **The quarterly review,** three months after implementation, s**howed 525% revenue growth** in transient business, with zero complaints about the online-only reservation policy.

2. What is creating friction that we should fix?

At *Synergy Dance Academy*, the **quarterly review revealed** that **parents loved online registration but were confused about costume ordering deadlines**. The system sent reminders, but parents wanted a visual countdown on the portal showing the number of days remaining to order.

3. What employee behaviors should we reinforce or correct?

AI creates new patterns, some good, some problematic. *Top Shop Auto* discovered that **mechanics loved the efficiency of text-based customer communication but had stopped calling customers about major repairs that require detailed explanations.** The review reinforced the rule: text for routine approvals, call for anything over $500, or safety-related.

Six-Month Performance Benchmarks

After six months, compare your business to the pre-AI baseline:

- ✓ **Customer volume:** Are you serving more customers without adding staff?

- ✓ **Revenue per customer:** Are approval rates and upsells improving?
- ✓ **Employee satisfaction:** Are skilled employees spending more time on expertise-driven work?
- ✓ **Customer retention:** Are customers coming back more frequently?
- ✓ **No-show rates:** Have automated reminders reduced missed appointments?

If any metric moved in the wrong direction, investigate immediately. AI should improve all of these metrics—if it is not, something in your implementation needs adjustment.

The Service Advantage

The most successful service businesses do not treat AI as a threat to the personal relationships that built their customer base or a first step toward becoming the corporate chains they compete against. **They use AI as an operations amplifier that requires experienced service providers** to fully realize its potential.

Key Success Elements:

- AI handles appointment scheduling, customer communications, routine inquiries, and follow-up reminders.
- **Service professionals** provide technical expertise, problem-solving, personal attention, and the community relationships that **turn first-time customers into lifelong regulars.**

- **Customer relationships remain personal** while service capabilities expand to match modern expectations.

- Quality stays high because **AI handles the phone calls** and paperwork that used to distract from the actual work.

Service business excellence is not about choosing between personal touch and operational efficiency. **It is about using AI to make experienced service providers more accessible, more responsive, and more valuable than ever before.**

Service businesses using AI create advantages that corporate chains and franchise operations cannot replicate. **The combination of instant responsiveness, personalized service, and genuine expertise takes years to develop.** A chain shop is just a transaction. A mechanic who remembers your car's quirks, who explains what is actually wrong in terms you understand, who calls you before the problem becomes expensive? That is service expertise no call center can fake.

Top Shop Auto did not lose their personal touch when they implemented AI phone systems. **They stopped missing calls during busy shop hours and started having real conversations with the customers** who walked through their door. *Synergy Dance Academy* did not become corporate when they automated registration. They eliminated the enrollment friction that was costing them students and freed up staff to focus on the families already in their programs. *MacDougalls' Marine Service* did not sacrifice its reputation when it added digital scheduling. They stopped losing customers to voicemail and started filling service slots that used to stay empty. *Salon 77 West* did not become impersonal when they automated appointment reminders. Their stylists stopped playing phone tag and started building the client relationships that drive referrals.

Here is what should keep service business owners who wait to implement AI up at night: While you are debating whether technology fits your neighborhood business philosophy, the shop across town just responded to that inquiry in two minutes. You responded in two hours. By the time you called back, they had already booked the appointment.

The auto repair shop that implemented AI scheduling six months ago never puts customers on hold during the morning rush. The dance studio that automated registration fills classes while you are still processing paper forms. The pet boarding facility that sends photo updates all day has customers bragging about them on social media. The marina that confirms reservations instantly fills slips that used to sit empty because someone forgot to return a call. **They are not becoming corporate chains. They are service businesses that decided "I will call you back" was no longer acceptable when customers expect instant answers from everyone else.**

Your customers have options. The franchise operations keep expanding. The app-based services keep improving. The corporate competitors keep getting more convenient. **The relationships you have built over years matter, but relationships do not survive indefinitely when responsiveness and convenience fall behind what customers can get elsewhere.** The loyal customer who has trusted you for a decade will still try the competitor who answered the phone on the first ring.

The customer expectations are not reverting to what they were before smartphones. The competitors are not going to stop improving. The corporate chains are not going to become less convenient. **The only question is whether you will use AI to become the service provider who combines genuine expertise with the responsiveness that modern customers demand, or whether you will keep telling**

yourself that quality alone will protect you from competitors who offer quality plus everything else.

Your technical skills are not being automated. Your community relationships are not being outsourced. Your ability to make a customer feel known and valued from the first phone call is not something a franchise script can replicate. But the service businesses that wrap that expertise in AI-enhanced operations will book more appointments. They will earn more referrals. They will build the neighborhood institutions that thrive while everyone else wonders where the customers went.

Here is what the businesses in this chapter understood that their competitors missed: **AI does not make you corporate. It makes you competitive.** It gives you the operational efficiency of larger competitors while preserving the personal service that customers actually prefer. The worry that technology will make your business "feel corporate" is backwards. What makes businesses feel corporate is when customers cannot get through, when messages go unanswered, when scheduling requires three phone calls. AI eliminates the friction. The personal touch was always up to you.

Top Shop Auto, Synergy Dance Academy, Edgartown Harbor, MacDougalls' Marine Service, Salon 77 West, Mission Auto KC—they all succeeded because they understood this distinction. **They did not implement AI to replace themselves. They implemented it to become the best version of themselves.** Technically excellent and operationally efficient. Personally connected and digitally convenient. Community-focused and professionally managed.

Their competitors watched from the sidelines, worried that AI would compromise their values. **And while they worried, they lost customers to businesses that proved you can be both high-tech and high-touch.**

How long before your best customers discover that the business down the street somehow always picks up the phone?

Chapter 17

Retail & E-commerce

Making Personal Service More Powerful

Your **retail customers expect personal service and technology, and they are willing to pay for it.** When John Modi, *MakerFlo* Founder, tells people his crafting company looks like "a million-dollar operation," he is not bragging—he is explaining how **AI turned his small team into** what customers believe is **a major enterprise.**

Retail businesses are adapting to customers' increasing expectations for the convenience often found in online shopping, such as immediate product availability, personalized recommendations, and efficient checkout processes.

Many **retailers continue to incorporate human assistance and community elements to support customer loyalty**. Rather than directly competing with large e-commerce platforms, some retailers seek to offer value through a combination of technology and interpersonal connection that differs from typical online-only models.

Success Story
Arizona Gifts

*Elena Rodriguez** built her family's gift shop into a beloved community institution over the course of 18 years. Customers chose *Rodriguez Gifts* not just for the products, but also for Elena's ability, accumulated over the years to suggest perfect gifts based on her extensive knowledge of families, relationships, and personal preferences. She remembered that Mrs. Johnson's grandson loved dinosaurs, that the Martinez family celebrated anniversaries with handwritten notes, and that teenage customers preferred trendy brands over traditional options.

The retail challenge: Customers increasingly expect the convenience that online shopping provides—instant product availability information, personalized recommendations, and seamless checkout—**while still valuing the human guidance and community connection that build their loyalty.**

Instead of viewing AI as a threat to her personal service advantage, Elena decided to **test whether technology could help her remember even more customer details and provide better recommendations** to a wider audience without compromising the warmth that defined her business.

Complete Customer Experience Integration

AI Analysis: System tracks customer Elena Martinez's 2-year purchase history, identifying preferences for handcrafted items, earth-tone colors, and quarterly gift-giving patterns for family celebrations.

Human Expertise: Store owner Elena applies AI insights within the context of personal relationship knowledge. Elena Martinez mentioned her daughter's upcoming quinceañera and has been planning celebrations for 6 months.

Enhanced Result: Perfectly timed quinceañera gift recommendations combined with data-driven personalization, with cultural understanding and family relationship knowledge. This resulted in a $340 sale and referral to three family friends.

Trust: Building Customer and Employee Confidence

Retail AI implementation faces unique challenges because customers directly experience technological changes through their shopping interactions, and employees worry about job security in an industry where fears of automation are widespread.

Scripts for the Customer Experience Evolution

For customers concerned about losing personal service: "AI helps me remember your preferences and family occasions better, so I can provide even more personalized recommendations when you visit."

For employees worried about job replacement: "This technology handles inventory tracking and checkout details so you can spend more time helping customers find exactly what they need and building relationships that create loyalty."

For community members questioning the future market leadership: "We are using smart systems to serve you better—with the same family values and personal attention that have defined our store for nearly two decades."

Ready: Preparing for Enhanced Customer Experience

Retail is unforgiving. Your customers compare you to Amazon whether you like it or not, and they make snap judgments about your business every time they walk through your door or land on your website. Before you install a single AI tool, you need to answer one question honestly: **Do you actually know your customers, or do you just think you do?**

Retailers with twenty years of experience swear they know their regulars inside and out. Then they try to set up a customer loyalty system and realize they have no purchase history, no email addresses, and no idea which customers spend the most. **All that knowledge lives in one person's head, and it is not backed up anywhere.** That is not a customer relationship; that is a liability waiting to happen.

Here is the preparation work that separates successful retail AI from expensive disappointment. First, audit your point-of-sale data. How far back does it go? Can you pull a report showing your top fifty customers by revenue? **If your POS cannot answer basic questions about buying patterns, you either need to upgrade your system or accept that any AI recommendation engine will be flying blind.** Second, look at your inventory tracking. AI cannot predict what to reorder if your stock counts are off by thirty percent because nobody updates the system after a busy Saturday. **Garbage in, garbage out—that rule has not changed since the first - computer was built.**

For e-commerce operators, the homework is different but just as critical. Your website analytics probably contain gold you have never mined. How many visitors abandon their carts? At what price point? On which products? **Most small e-commerce businesses have never looked at this data seriously because it felt like a project for "later." Later is now.** AI personalization tools need clean behavioral data to work their magic, and that means understanding what you already have before you pay someone to analyze it.

The retailers who thrive with AI are the ones who treat their data like inventory—something to be

counted, organized, and maintained. If you would not run your stockroom the way you run your customer database, you have work to do before any software demo.

Advance: Putting AI to Work Across Retail Types

Success Story *MakerFlo*

John Modi's crafting supply company demonstrates how small retailers can achieve enterprise-level results through strategic AI implementation.

***MakerFlo* achieved 8-figure revenue with an 8.5% conversion rate in just one year.** However, it is particularly noteworthy that they achieved this while maintaining the level of personalized service valued by customers, as evidenced by their over 1,000 customer ratings averaging 4.8 stars.

Implementation Strategy:

- ***Shopify Plus*** offers AI-driven recommendations to boost e-commerce performance.
- Implement AI tools to analyze customer behavior and personalization.
- Automation of social media for community building and customer engagement.

90-Day Results:

- AI-driven customer experiences increased revenue to eight figures.
- **Conversion rates improved to 8.5%** (industry average 2-3%).
- Maintained customer satisfaction at 4.8 stars throughout rapid growth.
- **Increased team productivity by 78%** while preserving personal service quality.

Success Story
CarGari Car Rental

Rafael Small established *CarGari*, a peer-to-peer car rental platform based in Colorado, successfully entering the market despite limited resources and competition from major industry players such as Turo. His local startup needed to deliver professional service and build customer trust while operating 24/7 with a lean team.

Rafael's solution was not to outspend the big players. He combined AI with local expertise and personal service. As he explains, "Google Ads are integral to *CarGari* for awareness of my services and my brand. The dashboard is user-friendly, and the data is easy to navigate. **Understanding how and where traffic is entering the platform allows me to adjust my strategy."**

The AI Tools That Made the Difference:

- ***Google Ads Performance Max*** campaigns with AI - for customer acquisition and brand awareness.
- ***Google Analytics 4* with AI insights** - for understanding customer behavior and optimizing marketing strategy.
- ***AI-powered Product Studio*** - for transforming simple photos into professional social media content.

Rafael's Results:

- 392 rentals achieved, projecting 2,180 by year-end.
- A 30% increase in rentals with 47% repeat customers.
- A 13% increase in vehicle owner sign-ups year over year.
- The local-focused approach was validated by 72% of customers arriving via Denver airport.

Rafael's insight:

"We do not try to be Turo. We try to be the best local option with technology that makes us reliable and professional."

Three Best AI Programs for Small Business Retailers

1. *Shopify Plus with AI* Recommendations - The Complete Retail Brain

What it does: AI-powered customer behavior analysis, automated inventory management, personalized product recommendations, abandoned cart recovery, and multi-channel selling both online and in-store.

Why retailers choose it: ***Shopify*** is designed specifically for businesses selling online and in-store, with effortless integration between digital and in-person services.

Real impact: *MakerFlo* achieved an 8.5% conversion rate and 8-figure revenues from an AI-enhanced customer experience—all with 4.8-star satisfaction ratings.

Monthly cost: $2,300-2,500+ per month base fee (depending on contract term and sales volume).

Best for: Growing retailers who need professional e-commerce capabilities integrated with in-store personal service.

2. Square for Retail with AI Analytics - The Customer Intelligence Engine

What it does: Smart inventory tracking, customer purchase pattern analysis, automated reorder suggestions, staff performance insights, and loyalty program management.

Why small retailers adopt it: ***Square*** provides enterprise-level customer intelligence for small businesses without the need for complicated setup or expensive implementation.

Real impact: Retailers using ***Square's AI*** features report a 34% improvement in inventory turnover, along with a 28% increase in customer return visits, thanks to better product availability and personalized service.

Monthly cost: $60-$165, depending on the number of locations and features.

Best for: Small to medium retailers who want customer intelligence without technical complexity.

3. Lightspeed Retail with AI Inventory - The Operations Optimizer

What it does: Predictive inventory management, sales forecasting, customer segmentation, staff scheduling optimization, and multi-location coordination.

Why growing retailers choose it: Lightspeed strikes a balance between sophisticated AI capabilities and user-friendly interfaces designed for retail staff.

Real impact: Multi-location retailers report a 45% reduction in stockouts and a 31% improvement in staff productivity through AI-optimized operations.

Monthly cost: $69-$229 per location, depending on the features.

Best for: Multi-location retailers and growing businesses that need comprehensive operational optimization.

Success Story
Sodapup

SodaPup, a dog toy manufacturer, effectively **employs AI-driven manufacturing optimization alongside direct-to-consumer e-commerce strategies.** This approach enhances operational efficiency while preserving distinct brand identity and fostering customer engagement that larger pet industry competitors find challenging to emulate.

Connect: Making AI and Personal Service Work Together

The magic happens when AI-powered customer insights work hand-in-hand with genuine human relationship knowledge to unlock new possibilities for personalized shopping experiences. Imagine receiving tailored product recommendations based on your preferences, paired with empathetic support from staff who understand your unique needs. These moments transform customer journeys, creating connections and value that neither technology nor traditional service could achieve alone.

Success Story *Thriftbooks*

ThriftBooks shows how **technology can strengthen, rather than replace, community ties.** The company utilizes AI-driven inventory management to ensure stores have the right books available and **leverages intelligent customer recommendation systems to personalize each shopper's experience.**

Alongside these technological advances, *ThriftBooks* **actively supports local communities** through book donation programs and literacy initiatives. This approach demonstrates how retailers can utilize advanced tools to simultaneously optimize business operations and cultivate stronger relationships with the neighborhoods they serve.

Keep: Maintaining Retail Excellence

Retail businesses that successfully implement AI create advantages that strengthen over time by establishing systems that enhance customer relationships while improving operational performance.

Scaling What Works

Review your first six months of data. **Which AI insights led to your biggest improvements in customer satisfaction or sales conversion?** Double down on those areas. If personalized recommendations increased average order value by 28%, expand that system to cover more product categories. If inventory optimization reduced stockouts by 45%, implement it across all locations.

Many retailers make the mistake of adding new AI tools instead of optimizing the ones they already have. Resist this temptation. Perfect your current systems before expanding.

Avoiding the Success Trap

Success can create overconfidence. Do not start automating everything just because your first AI tools worked well. Customers still value the personal attention and community connection that only local businesses can provide. AI should enhance these relationships, not replace them.

Monitor both sales metrics and customer satisfaction scores closely. If customer loyalty starts declining or community engagement decreases, investigate whether technology is interfering with the personal service that built your reputation.

The Retail Advantage

Retail businesses that effectively leverage **AI enhancements establish competitive advantages** that not only endure but also intensify over time. Unlike pure e-commerce platforms that compete primarily on price and convenience, AI-enhanced local retailers can offer unique values that online-only businesses cannot imitate:

AI enables comprehensive customer knowledge, and **the most successful retailers do not treat AI as a replacement for the personal service that built their customer loyalty** or an attempt to 'out-Amazon' Amazon. They use it as a business amplifier that requires experienced retailers to fully realize its potential.

Key Success Elements:

- **AI handles inventory prediction,** customer data management, personalized marketing, and routine communications.
- **Retailers provide** curated selection, expert product knowledge, community connection, and the **personal relationships that turn shoppers into regulars.**
- Customer relationships remain personal while service capabilities expand to compete with digital convenience.
- Product expertise stays sharp because **AI handles the logistics that used to distract from actually helping customers.**

Retail excellence is not about choosing between personal service and digital efficiency. **It is about using AI to make experienced retailers more knowledgeable, more responsive, and more valuable than ever before.**

Retailers thoughtfully using AI create advantages that e-commerce giants and big-box chains cannot replicate. **The combination of instant inventory intelligence, personalized outreach, and genuine human connection**

takes years to develop. An Amazon recommendation is just an algorithm. A shop owner who remembers your daughter's birthday is next month, who knows you prefer vintage over modern, who sets aside the perfect piece before you even know you want it? That is retail expertise no platform can provide.

The boutique owner who implemented AI inventory management did not lose her curatorial eye. She stopped running out of bestsellers on Saturday afternoons and started having the buying budget for the unique pieces her customers actually come in for. The sporting goods store that added customer data tracking did not become impersonal. They started reaching out to fishing customers when the seasonal gear arrived instead of hoping people would wander in at the right time. The gift shop that automated personalized marketing did not sacrifice their community feel. They remembered every customer's preferences and sent birthday reminders that felt thoughtful, not corporate.

Here is what should keep retailers who wait to implement AI up at night: While you are debating whether technology fits your small-town shop philosophy, the store three blocks over just sent personalized emails to every customer who bought holiday gifts last year. They know exactly what is selling and what is collecting dust. Their reorder system caught the inventory gap before the weekend rush. Their loyalty program remembers purchase history you forgot existed.

The retailer that implemented AI customer intelligence six months ago greets regulars by name and knows what they bought last time. The e-commerce shop that added personalized recommendations sees cart values climb because the suggestions actually make sense. The specialty store that automated inventory tracking never apologizes for being out of stock on the item customers drove across town to find. **They are not selling out to big-box tactics. They are shop**

owners who decided that "let me check in the back" was no longer acceptable when customers can find anything online in thirty seconds.

Your customers have options. The e-commerce platforms keep getting faster. The big-box stores keep expanding selection. The direct-to-consumer brands keep multiplying. **The relationships you have built over years matter, but relationships do not survive indefinitely when product availability and personalized service fall behind what customers can get with a few clicks.** The loyal customer who has shopped with you for a decade will still try the competitor who remembered their size, their style, and their budget without being asked.

The retail landscape is not reverting to what it was before smartphones. The customer expectations are not getting more patient. The online competitors are not going to stop improving. **The only question is whether you will use AI to become the retailer who combines genuine personal service with the intelligence that modern shoppers demand**, or whether you will keep telling yourself that charm alone will protect you from competitors who offer charm plus everything else.

Your product expertise is not being automated. Your community relationships are not being outsourced. Your ability to make a customer feel known and valued the moment they walk through your door is not something a website can replicate. But the retailers who wrap that personal touch in AI-enhanced operations will move more inventory. They will build stronger loyalty. They will become the neighborhood institutions that thrive while everyone else blames Amazon for empty aisles.

Here is what the successful independent retailers understand that struggling ones often miss: **you are not competing with Amazon on selection or price. You never were.** You are competing on expertise, on curation, on the experience of walking into a space where someone actually knows what they are talking about and cares whether you find the right thing. AI does not threaten that advantage. It amplifies it by giving you customer knowledge and operational efficiency that lets you focus on what you actually do better than any algorithm.

The retailers who understand this are not becoming less personal. **They are becoming the kind of shop they always wanted to be, with AI handling the inventory headaches and customer data so they can focus on the relationships that made them open their doors in the first place.**

How long before your best customers start discovering that the shop down the street somehow always has exactly what they need?

Chapter 18

Healthcare & Wellness

Combining Technology with Care

Wellness professionals who have not yet adopted AI are denying their patients what they increasingly expect: personalized care and attention that technology can provide. From **remembering health goals to tracking progress** and ensuring nothing is overlooked. While still highlighting the unique healing relationships and expertise of skilled practitioners.

Success Story
Community Psychiatry

*Dr. Sarah Che** runs a small mental health practice in Portland, Oregon, serving eighty-five regular clients with a team of four therapists. For 12 years, **she had established a reputation for providing personalized care and comprehensive treatment plans**. But, like every therapist, Sarah was overwhelmed by the administrative work that kept her away from patients.

Dr. Che dedicated two to three hours each evening to preparing progress notes, routinely **working until 10 PM to ensure her documentation complied with insurance standards.** Her therapists were burning out from the same issue—great at therapy but drained by documentation.

When one of her senior therapists announced he was leaving because **"I went into therapy to help people, not to fight with insurance paperwork,"** Sarah knew something had to change.

Instead of viewing AI as a threat to therapeutic relationships, Sarah decided to test ***SimplePractice AI Note Taker*** to handle the documentation burden that was stealing time from patient care.

Trust: Getting Buy-In from Therapists and Patients

The conversation with therapists and patients about AI use in clinical practice requires careful positioning to preserve the healing focus that defines successful therapy.

Scripts for Protecting the Therapeutic Relationship

For the therapist concerned about AI interfering with therapy: "The note-taking system runs quietly in the background. You focus completely on your client, just like you always have. After the session, you get a draft of progress notes based on what actually happened, which you review and approve. Your therapeutic skills become more powerful, not less important."

For the therapist concerned about accuracy: "You still review and approve every note. The AI just gives you a solid first draft based on what actually happened during the session, so when you begin making your session notes, you are not starting from a blank page."

For the therapist worried about client privacy: "***SimplePractice*** is HIPAA-compliant with the highest security standards. The AI processes information to create notes, but you control what information is included and how the data is used."

Getting Your Team Prepared

Before you choose specific AI tools for your practice, your practitioners need to be confident about the changes. Start with individual conversations explaining how AI documentation systems will eliminate the charting that exhausts everyone, giving them more energy for the healing work they love.

Schedule team meetings to demonstrate the technology. Let therapists ask questions about job security, training requirements, and how their daily routines will change. **Address concerns head-on instead of dismissing them.**

Demonstrate how AI can manage routine documentation, allowing them to focus on therapeutic relationships and complex case management. **Position the technology as protecting healing work, not changing it.**

Scripts for Building Client Trust

The conversation with patients about AI in wellness practices should follow a specific pattern:

For Therapy Patients: "Our new documentation system helps me stay more present during our sessions and remember important details from our previous meetings, so I can focus entirely on supporting your progress."

For Wellness Coaching: "The progress tracking tools help me understand patterns in your wellness journey and suggest approaches that have worked for similar goals—without replacing our personal coaching relationship."

For Mental Health Treatment: "The treatment planning assistance helps ensure consistency in your care and tracks your progress toward goals, but all therapeutic decisions and insights come from our work together."

It is essential to emphasize that AI serves as a tool to enhance the attentiveness and consistency of human care, rather than a substitute for it.

Ready: Preparing Systems & People

Wellness practice AI implementation succeeds when it reinforces, rather than competes with, **healing expertise and patient relationships.** Select tools that streamline administrative tasks, allowing you to focus on therapy and care.

Healthcare has compliance requirements that can turn a simple software purchase into a regulatory nightmare. Before you get excited about any tool that promises to revolutionize your practice, you need to know exactly what you are getting into.

HIPAA is not optional, and "HIPAA-compliant" on a vendor's website does not mean your implementation will be compliant. We have watched wellness practitioners sign up for AI documentation tools without understanding that they are now responsible for how that data is stored, transmitted, and protected. The vendor handles their end, but your Business Associate Agreement, your breach notification procedures, and your staff training on data handling are all on you. **If your current compliance documentation is a dusty binder from five years ago, updating it comes before any AI purchase.**

The preparation that actually matters starts with understanding your documentation burden. Pull out a stopwatch for one week and track how long your practitioners spend on notes, insurance coding, treatment plans, and appointment scheduling. **We are not talking about estimates. We mean real numbers.** When a therapist tells us she spends "a couple of hours" on paperwork each day, and then we actually measure it, the number is usually closer to four hours. **You cannot fix what you have not measured, and you cannot evaluate whether an AI tool is worth $2,000 a month if you do not know how many hours it needs to save to pay for itself.**

Next, look at your current systems honestly. Can your practice management software integrate with AI tools, or are you running something from 2012 that requires manual data entry for everything? How clean is your patient data? Are your treatment notes standardized enough that an AI tool could learn patterns from them, or does every practitioner document differently? **These questions matter because AI tools learn from your existing information. If your records are a mess, the AI will learn from that mess.**

Lastly, and this is where many wellness practices fall short, **consult your malpractice insurance provider before implementing any changes.** Some carriers have specific requirements about AI-assisted documentation and clinical decision support. **Finding out your coverage has gaps after you have been using a tool for six months is not a conversation you want to have.**

The practices that succeed with AI are the ones that treat preparation like patient intake: thorough, well-documented, and honest about the overall system wellness.

Mental Health Practice Implementation

Dr. Che's 90-day plan shows wellness methods can boost efficiency and improve patient care.

30-Day Foundation Period:

- ***SimplePractice AI Note Taker*** was deployed across all four therapists.
- **Documenting time decreased** from 15-20 hours weekly to 3-4 hours for the entire practice.
- Therapists may choose to **maintain eye contact during sessions** rather than take notes.
- The **complete elimination of evening work** throughout the entire practice not only improved work-life balance for the team but also enhanced overall productivity and staff satisfaction.

60-Day Capability Expansion:

- **Implemented patient progress tracking** with optimized treatment plans.
- Automated appointment reminders with enhanced follow-up communication.
- Established quality control for AI documentation.

90-Day Growth Phase Results:

- Treatment plan consistency improved among all practitioners.
- Client no-show rates decreased by 23% through automated reminders.
- Insurance claim accuracy increased due to improved documentation.

6-Month Results:

- Overall practice efficiency improved 67% through AI-human collaboration.
- Patient satisfaction increased to 94% due to improved therapist interaction.
- Practitioner burnout was resolved, with all therapists maintaining standard working hours.
- Revenue grew by 34% as a result of enhanced operational capacity and higher patient retention rates.

Three Best AI Programs for Health and Wellness Practices

1. SimplePractice with AI Note Taker - The Complete Wellness Practice Brain

What it does: AI-powered clinical documentation that generates progress notes from session recordings or summaries. Integrates intelligent appointment optimization, patient progress tracking with insights, and automated treatment plan reminders, all with HIPAA-compliant patient communication.

Why wellness practitioners choose it: ***SimplePractice*** is designed specifically for mental health therapists, massage therapists, nutritionists, and wellness practitioners with industry-specific workflows and compliance requirements.

Real impact: Dr. Matthew Hitchcock reported that AI assistance reduced his documentation time from two hours to just twenty minutes per case six times efficiency improvement. Mental health professionals are experiencing similar benefits.

Monthly cost: $49-99, depending on the practitioner count, with **AI Note Taker** available for an additional $35 per practitioner.

Best for: Licensed wellness practitioners who need comprehensive practice management with AI-enhanced clinical documentation.

2. Mindbody with AI Customer Experience Management - The Client Engagement and Retention Powerhouse

This tool is listed twice in this book – it is that good! **What it does:** Messenger AI offers a 24/7 automated assistant for client inquiries, predictive analysis to spot at-risk patients, class and service optimization using attendance data, and personalized wellness journey mapping.

Why wellness businesses are adopting it: Mindbody's "Clients At-Risk" feature utilizes machine learning to predict when clients are losing motivation, alerting practices to nurture these relationships.

Real impact: Practices using the "Clients at Risk" feature are seeing 30% of those clients become more engaged after implementing the tool. One practitioner reported: "In the first three months, I DOUBLED my income with ***Mindbody***."

Monthly cost: $129-399, depending on the features and business size.

Best for: Fitness studios, wellness centers, and practices focused on building client communities and long-term wellness relationships.

3. AI Clinical Documentation Tools - The Treatment Excellence and Compliance Assistant

Multiple platforms serve clinical documentation, including ***Blueprint AI, Mentalyc***, and ***Quill***, which transform clinical documentation:

What they do: AI-powered progress note generation, real-time session transcription, treatment plan automation, and clinical insights that identify progress patterns.

Why serious practitioners choose them: AI therapy notes have reduced paperwork time by 40% and improved the quality of care, helping therapists focus more on their patients by simplifying documentation—all HIPAA-compliant.

Real practitioner feedback: "I write a free-form account of the session content, then ask the AI to format it into a proper DAP or SOAP note. It does a remarkably good job, and I just review and edit the final product."

Monthly cost: $30-$ 80 per practitioner, depending on the platform and features.

Best for: Licensed mental health professionals who need specialized clinical documentation support.

Connect: Making AI and Healing Expertise Work Together

True **success in wellness practices is achieved when AI's analytical capabilities**, such as providing personalized data insights and identifying patterns in client progress, **work hand in hand with human therapeutic expertise** while including empathy, intuition, and clinical judgment. AI can review session notes and recommend tailored interventions based on each client's unique needs, while therapists bring warmth, understanding, and a compassionate approach to care.

For example, AI might suggest a mindfulness exercise for a client showing signs of stress; the therapist then delivers the intervention with genuine support and adapts it to the client's personal circumstances. This collaboration enables more effective, individualized healing experiences and leads to better outcomes for clients.

The Complete Workflow Integration Pattern

The strongest wellness practice AI implementations consistently demonstrate similar success:

AI Handles: Session documentation, appointment scheduling, client communication, progress tracking, insurance billing, and administrative tasks.

Human Expertise Provides: Therapeutic relationships, clinical judgment, treatment planning, crisis intervention, a healing presence, and cultural sensitivity.

Working Together Results: The team boasted of more present practitioners, consistent documentation, personalized care at scale, and protected healing relationships with patients.

Real-World Integration Success

AI Documentation Integration:

AI Documentation: The system records session details and generates initial progress notes based on therapeutic content.

Human Interpretation: The therapist reviews the notes within the therapeutic context, adds clinical insights, and ensures accuracy for effective treatment planning.

Working together results: Therapists maintain complete focus on the client during sessions while ensuring thorough documentation for continuity of care.

Personalized Wellness Journeys:

AI Analysis: ***Mindbody's*** machine learning identifies client engagement patterns and predicts motivation levels to prevent therapy dropouts.

Human Connection: Wellness practitioners utilize AI insights to tailor their approaches and deliver timely support personalized to individual patient needs.

Collaborative Outcome: Higher patient retention through proactive intervention combined with personalized therapeutic relationships.

Real Practitioner Testimonials

Real therapist testimonials about ***Blueprint AI*** show the impact: "***Blueprint*** saves me a lot of time, helps me be more present with my patients, and the notes are solid as well as insurance compliant. I wish I had it sooner!"

Another practitioner reported: "I could not be happier with ***Blueprint***. It has allowed me to go from a part-time therapist to a full-time therapist, all while maintaining a good work-life balance."

Real-World Trust Building Results

Dr. Che's implementation exceeded her expectations:

Practitioner Response:

- 100% of therapists reported **feeling more present** during sessions.
- **Documentation quality improved** because AI previously missed details.

- **Eliminated recordkeeping burnout**, and everyone went home on time.
- Job satisfaction increased as **therapists could focus on healing work.**

Patient Response:

- 94% of **patients noticed improved therapist attention** during sessions.
- **Patients appreciated therapists remembering details** from previous sessions.
- **Patients did not have concerns about privacy** or technology interfering with care.
- Several patients commented that their **therapists seemed "more focused and organized."**

This pattern is consistent for all wellness practices: framing AI as supportive rather than a substitute for human care leads to positive responses from both practitioners and patients.

Keep: Maintaining the Momentum

AI-powered wellness **practices gain lasting advantages by improving healing systems** and maintaining their core therapeutic culture.

Scaling What Works

Which AI insights led to your biggest improvements in patient outcomes? Increase your efforts in that area. If AI documentation reduced evening work from 15 hours to 2 hours weekly and improved therapist job satisfaction by 40%, expand that system to track treatment progress patterns and clinical insights. If predictive client retention prevented twenty dropouts and generated $15,000 in retained revenue, create more detailed engagement tracking and intervention protocols.

Many wellness practices make the mistake of adding new AI tools instead of mastering the ones they already have. Resist this temptation. **Perfect your current systems before expanding.**

Clinical Documentation Innovation

Dr. Matthew Hitchcock's experience demonstrates the transformative power of AI-assisted clinical documentation. Hitchcock reduced documentation time from two hours to twenty minutes per case while improving care quality through better focus on patient interaction.

The Wellness Practice Advantage

The most successful wellness practitioners do not treat AI as a threat to their therapeutic relationships or a shortcut around the healing presence that patients need. They **use it as a practice amplifier that requires experienced clinicians to fully realize its potential.**

Key Success Elements:

- AI handles documentation, scheduling, insurance coding, and routine patient communications.

- **Practitioners provide therapeutic presence**, clinical judgment, emotional attunement, and the healing relationships that drive outcomes.
- **Patient relationships remain personal** while practice capabilities expand to serve more people who need help.
- **Compliance stays tight** because AI handles the administrative burden that used to steal time from actual care.

Excellence is not about choosing between healing presence and operational efficiency. **It is about using AI to make experienced practitioners more present, more focused, and more effective than ever before.**

Wellness practices thoughtfully using AI create advantages that corporate healthcare chains and telehealth platforms cannot replicate. **The combination of instant responsiveness, comprehensive documentation, and genuine therapeutic connection takes years to develop.** A telehealth app is just a transaction. A therapist who remembers the breakthrough you had six sessions ago, who notices the subtle shift in your voice that signals something is wrong, who creates a space where healing actually happens? That is clinical expertise no algorithm can provide.

The physical therapist who implemented AI documentation did not lose her clinical eye. She stopped spending evenings typing notes and started being fully present with the patient in front of her. **The mental health practice that added AI scheduling did not become impersonal.**

They stopped losing patients to phone tag and started filling the cancellation slots that used to stay empty. The chiropractor who automated insurance coding did not sacrifice patient care. He stopped fighting with billing systems and started having the conversations about lifestyle and prevention that patients actually need.

Here is what should keep wellness practitioners who wait to implement AI up at night: While you are debating whether technology belongs in a healing environment, the practice across town just eliminated their documentation backlog entirely. Their clinicians leave at 5 PM instead of 8 PM. Their patients get appointment reminders that actually reduce no-shows. Their insurance claims go out clean the first time.

The therapist who implemented AI note-taking six months ago spends sessions fully present instead of half-listening while typing. The acupuncturist who automated patient intake gathers comprehensive health histories before appointments instead of rushing through forms. The counseling practice that added AI-assisted treatment tracking identifies patterns that would take months to notice manually. **They are not compromising their healing philosophy. They are practitioners who decided that administrative drowning was no longer acceptable when their patients needed them fully present.**

Your patients have options. The corporate wellness chains keep expanding. The telehealth platforms keep getting more convenient. The insurance networks keep adding providers. **The therapeutic relationships you have built over years matter, but relationships do not survive indefinitely when appointment availability and communication responsiveness fall behind what**

patients can get elsewhere. The client who trusts you with their mental health will still try the new practice that actually answered the phone when they were in crisis.

The documentation requirements are not getting simpler. The insurance coding is not getting easier. The patient expectations are not reverting to what they were before instant everything. **The only question is whether you will use AI to become the practitioner who combines genuine therapeutic presence with the accessibility that modern patients demand, or whether you will keep telling yourself that clinical skill alone will protect you from practices that offer clinical skill plus everything else.**

Your healing presence is not being automated. Your clinical intuition is not being outsourced. Your ability to create a space where people feel safe enough to do the hard work of getting better is not something software can replicate. But the practitioners who wrap that presence in AI-enhanced operations will see more patients. They will reduce their burnout. They will build the practices that thrive while everyone else wonders why they went into healthcare in the first place.

Here is the truth that wellness practitioners need to hear: **the administrative burden is not noble suffering. It is theft.** Every hour you spend on documentation after patients leave is an hour stolen from your family, your rest, your own wellbeing. Every evening lost to insurance coding is energy you cannot bring to tomorrow's sessions. AI does not diminish your healing work. It protects you from the system that was grinding you down.

The practitioners who understand this are not abandoning their therapeutic values. **They are reclaiming the reason they became healers in the first place, with AI handling the paperwork so they can focus on the people who walk through their door asking for help.**

How long before your patients start noticing that the other practitioner somehow always seems more present and less rushed?

Chapter 19

Restaurant & Food Service

AI That Enhances Hospitality

Your customers are not coming to your restaurant to talk to a computer—they are coming for great food and personal service.

When we tell restaurant owners that AI will help them serve customers better, we usually get the same worried response: "My customers come here for my grandmother's recipes and the fact that we remember their names. They do not want to order from a robot or eat food made by machines."

What some restaurant owners do not realize is that **customers now expect the convenience that technology provides.** This involves reducing waiting times, ensuring order accuracy, and providing individualized service, while also maintaining the personal touch involved in dining. Rather than presenting a choice between AI and genuine hospitality, the solution lies in leveraging AI to enhance the consistency of personalized service and improve operational efficiency.

Success Story
Poppy's Pizza & Grill

For three decades, *Poppy's Pizza & Grill* has been delighting diners with sensational Italian food at the entrance of the Rocky Mountains in Estes Park, Colorado. But the seasonal nature of their location created the same challenge that drives every restaurant owner crazy with phones ringing off the hook during busy periods while staff are trying to serve the customers standing right in front of them.

When staffing **shortages left those phones unanswered during peak tourist seasons,** *Poppy's* owner realized they were losing money every time a potential customer's call went unanswered, received a busy signal, or gave up after remaining on hold.

The restaurant's challenge: How do you provide the instant response and seamless ordering that customers now expect—without losing the warm hospitality and authentic food preparation that keeps them coming back?

Poppy's solution was not to replace human service with technology. **They used AI to handle routine order coordination so their staff could focus entirely on food preparation and in-person customer care.**

Trust: Building Confidence in AI-Enhanced Hospitality

Food service AI implementation faces unique challenges because customers directly experience technological changes through their dining interactions, and staff worry about job security in an industry where automation fears are widespread.

Hospitality Enhancement

Food service AI implementation is effective when it supports culinary skills and customer relations, without replacing them. AI drives efficiency, tracks diners' preferences, and streamlines communication to scale and standardize your hospitality.

Hospitality Principles:

Principle 1: AI Remembers; Humans Connect - Use AI to track customer preferences, dietary restrictions, and visit history, then apply human warmth to interpret these insights within personal and cultural contexts.

Principle 2: Technology Optimizes Operations; Relationships Drive Satisfaction - Use AI for inventory forecasting, order management, and kitchen coordination while using saved time for customer interaction and customizing diners' experiences.

Principle 3: Automation Handles Routine; Hospitality Focuses on Experience - Automate reservations, order processing, and payment handling to create more time for menu explanation and recommendations, special experience accommodations, and memorable service moments.

Principle 4: AI Expands Reach; Humans Preserve Culture - Use technology to serve more customers across in-restaurant, takeout, and catering. Maintain the authenticity and distinctive touch that define your restaurant's identity.

Scripts for Building Staff and Customer Confidence

For customers concerned about losing personal service: "AI helps us remember your preferences and family occasions better, so we can provide even better service when you visit."

For staff worried about job security: "This technology handles inventory tracking and order coordination so you can spend more time creating memorable dining experiences for our guests."

For community members questioning market position: "We are using smart systems to serve you better while maintaining the same family values and personal attention that have defined our restaurant."

Ready: Preparing for Enhanced Customer Experience

Restaurant owners have been burned by technology promises more times than they can count. That point-of-sale system that was supposed to be intuitive? Your servers still cannot figure it out two years later. The online ordering platform that was going to change everything? It takes a 15% cut of every order, and your customers complain that the food is cold by the time they pick it up. So, when someone tells you AI will transform your restaurant, **your skepticism is earned.**

Here is what you need to know before you spend a dime: **restaurant AI only works if your basic operations are already solid.** No software is going to fix a kitchen that cannot get orders out on time or a front-of-house team that forgets to greet customers. **AI amplifies what you already do. If what you already do is chaos, you are just going to produce faster chaos.**

Begin with your phone. How many calls go unanswered during a Friday dinner rush? If you do not know, you do not know how much money you are leaving on the table. Track it for two weeks. Count the rings that nobody picks up, the customers who hang up after being on hold, the voicemails that never get returned. **That number is your baseline, and it tells you exactly what an AI phone system needs to recover to justify its monthly fee.**

Next, look at your ticket times and order accuracy. Pull your POS data for ticket times (assuming your POS actually tracks them) and calculate your average ticket time. What percentage of orders go out wrong? Where do mistakes happen? At the order-taking stage, in the kitchen, or at handoff? **AI tools can help with all three issues, but you need to know which problem you are actually solving.** A restaurant that implements AI order-taking when its real issue is kitchen coordination has just moved the bottleneck somewhere else.

Of course, you cannot overlook inventory. Most independent restaurants run on gut instinct and whatever the owner remembers to order on Tuesday. AI inventory systems can predict demand and reduce waste, but they need data to learn from. **If your current inventory tracking is a clipboard and a prayer, you will spend the first three months just teaching the system what you actually use.** That is not wasted time; it is necessary groundwork. You cannot expect magic results in week one.

Finally, have an honest conversation with your staff before you roll anything out. **Restaurant workers have seen "efficiency tools" that really meant "we're cutting your hours."** If your servers think AI phone answering is the first step toward replacing them, they will sabotage it without even meaning to. **Tell them the truth: this handles the stuff that pulls them away from tables so they can focus on the work that actually earns tips.**

The restaurants that win with AI are the ones that know their numbers, fix their basics first, and bring their teams along for the ride.

Advance: Putting AI to Work

Restaurant and food service operations leverage AI for ordering systems and inventory management. AI technologies can also help optimize staff scheduling and predict customer preferences by analyzing historical purchase data. Additionally, these tools aid in reducing food waste and improving overall efficiency across various operations within the industry.

Success Story| *Poppy's Pizza & Grill*

Poppy's has implemented ***Popmenu AI Answering*** to handle phone calls when staff are unavailable. The decision to implement AI was not about eliminating people; it was about enabling them.

The Verified Results in 3 Months:

- The AI system fielded 3,600 calls (40 calls per day).
- **43% of the calls were tied directly to revenue activity.**
- **Generated over $76,000** in online orders.
- Staff could serve the customers standing in front of them instead of taking care of remote customers calling the business.

AI Implementation Strategy:

- ***Toast.ai*** order management was introduced to coordinate phone and online orders.
- AI inventory predictions reduced food waste and improved ingredient availability.
- Automated customer communication systems were implemented to handle order updates and manage loyalty programs.

Results after 6 months:

- Phone **order processing time dropped 67%**—with food order accuracy intact.
- **Food waste was decreased by 34%** through better food demand prediction.
- Due to reliable order communication, customer satisfaction increased by 78%.
- From expanded ordering capacity, **revenue grew by 89%.**

Three Best AI Programs for Restaurants

1. Toast with AI Order Management - The Complete Restaurant Brain

What it does: AI-powered menu optimization, inventory forecasting, staff scheduling, customer preference tracking, and seamless connections with your existing systems.

Why restaurant owners choose it: Toast was designed specifically for restaurants, featuring workflows for kitchen operations, customer service, and business management.

Real impact: Restaurants are most commonly using AI for optimizing menu performance through data analysis of ordering patterns and profitability metrics.

Monthly cost: $69-400+ per month for software (depending on features and staff count), plus $500-1,000 for hardware, and payment processing fees of 2.49-3.69%. Most full-service restaurants pay $200-400+ monthly when including add-ons like online ordering, kitchen display systems, and loyalty programs.

Best for: Full-service restaurants that need comprehensive operations management with AI optimization.

2. *Resy for Restaurants with AI Table Management - The Customer Experience Optimizer*

What it does: Intelligent reservation management, customer preference tracking, table optimization, waitlist automation, and dining pattern analysis.

Why hospitality professionals adopt it: Resy offers sophisticated reservation management that helps restaurants maximize capacity—without sacrificing personal service.

Real impact: Restaurants using ***Resy*** report a 34% improvement in table utilization and a 28% reduction in customer wait times, thanks to optimized seating and accurate predictions of table turn times.

Monthly cost: $249-899, depending on features and the number of diners served.

Best suited for: Upscale restaurants focused on reservation management and optimizing customer experience.

3. OpenTable with Guest Intelligence - The Relationship Building Engine

What it does: Comprehensive guest profiles, dining history tracking, preference management, automated marketing, and reservation optimization.

Why relationship-focused restaurants choose it: OpenTable combines reservation management with deep customer intelligence, enabling personalized service.

Real impact: Restaurants report a 45% increase in repeat visits and a 67% boost in customer satisfaction through personalized service based on dining history and preferences.

Monthly cost: $249+, depending on the number of diners and features.

Best for: Restaurants focused on building lasting customer relationships and encouraging repeat visits through personalized hospitality.

Connect: Making AI and Culinary Expertise Work Together

The magic occurs when AI-driven customer insights merge with human hospitality skills, crafting dining experiences that neither technology nor conventional service can accomplish independently.

Human-AI Collaboration Examples

Personalized Menu Recommendations:

AI Analysis: Mrs. Peterson's 18-month dining record shows she prefers lighter proteins, low-sodium meals, and often dines on Tuesday evenings.

Human Expertise: Server Maria interpreted AI insights within the context of Mrs. Peterson's recent mention of doctor-recommended dietary changes and an upcoming anniversary celebration.

Enhanced Result: A menu was suggested that aligns with health requirements and is suitable for celebrations, resulting in a $78 order.

Kitchen Workflow Optimization:

AI Coordination: The system analyzed order timing, ingredient prep requirements, and kitchen capacity to optimize cooking sequence during a busy Friday evening

Culinary Expertise: The Head chef used AI recommendations while applying his 15 years of experience to make ingredient substitutions and special customer requests.

Superior Outcome: Achieved 41% faster ticket times—food quality stayed high, and no customer complaints about waiting times.

Cultural Experience Enhancement:

AI Information: The system recognized new customers and supplied information regarding traditional family recipes and the history of the restaurant.

Human Warmth: Elena personally welcomed new guests, shared family traditions, and told stories about her grandmother's recipes, all while AI handled routine coordination.

Memorable Result: Authentic cultural experience enhanced by seamless service efficiency, leading to positive reviews and repeat visits.

What Could Go Wrong (And How to Prepare)

Implementing AI in restaurants seems straightforward in sales demos and case studies. The reality is often messier. Here is what actually happens.

When the AI phone system crashes during Friday dinner rush, hope is not a plan. “It will happen” as your backup plan cannot be "refresh the browser and pray."

Have a dedicated phone line that bypasses the AI system. Train your host staff to handle manual reservations and takeout orders without AI, even if it means using paper and pen. **Test these backup procedures** during a slow Tuesday lunch, not when you discover you need them at 7 PM on Valentine's Day.

When AI inventory predictions miss your seasonal spikes:

The algorithm does not know about the high school football championship, the sudden cold snap that triples soup orders, or that your town has three major conventions scheduled the same week. Your experienced kitchen manager knows these things. Give them authority to override AI ordering recommendations when their local knowledge tells them something is wrong. Keep manual ordering procedures current and accessible. Document these overrides to help the system learn your market's unique patterns.

When automated ordering creates kitchen confusion:

Online orders will come through with modifications that the AI did not properly flag. Table 12 will order the fish special that you ran out of 20 minutes ago because the inventory sync lagged. Your line cooks will hate the system if it creates more problems than it solves. Establish clear protocols: who confirms specialty orders with the kitchen before accepting them, how the system flags sold-out items in real-time, and when staff override AI decisions. Invest time upfront training both the technology and your team on your specific workflow.

When customers push back on "talking to a robot," Some longtime customers will resist. They want to hear a familiar voice when they call for reservations or takeout. **Have your AI system transfer seamlessly to staff for customers who prefer human interaction.** Program it to recognize repeat customers and offer the option: "Would you like me to connect you directly with Maria?" Make the technology optional, not mandatory. Your goal is convenience for customers who want it, not forcing everyone into one system.

The learning curve nobody mentions:

Months three through six are the danger zone. The novelty has worn off. Your kitchen staff is frustrated because the system does not work as smoothly as promised. Your servers are annoyed when the AI sends them conflicting table status updates. Customer complaints might actually increase compared to your old methods. This is when restaurants quit.

Push through. Schedule weekly check-ins focused on specific problems, not general complaints. Celebrate small wins. Track improvements that matter to individual staff members, not just overall restaurant metrics. Give your most skeptical server veto power over customer-facing AI features. When they finally say it is working, everyone else will believe it.

Keep: Maintaining Hospitality Excellence

Restaurant businesses that successfully implement AI create advantages that strengthen over time. Establishing systems that improve food operations without disrupting established hospitality practices.

Cultural Preservation with Modern Convenience

AI tools can assist with traditional recipes and service styles without imposing uniformity. Technology can expand experiences, cultural practices, and family customs intact. **Balancing automated systems with personal interaction preserves the emotional connections** fundamental to hospitality.

Real-World Trust Building Results

The restaurant's implementation exceeded expectations:

Staff Response:

- 96% of the **restaurant staff expressed confidence in AI** implementation plans.
- Kitchen staff **recognized AI as a tool for highlighting their culinary expertise** more effectively.
- Servers valued improved **abilities to build customer relationships.**
- Management appreciated the decreased demands related to coordination and scheduling.

Customer Response:

- There was a **92% positive reaction** to service enhancements.
- Customers valued clear communication about better accuracy and menu options.
- **No concerns were raised** about losing cultural authenticity or personal service.
- Several **customers requested information about private dining** and catering services.

This trend of enhanced performance resulting from **AI implementation is evident across all categories of restaurants.** When you position AI as enhancing, rather than replacing, hospitality, both staff and customers respond positively.

The Food Service Advantage

Successful restaurant owners view AI as a tool to enhance operations, not as a threat or replacement for genuine hospitality, relying on their expertise to maximize its benefits.

Key Success Elements:

- **AI handles phone answering, reservation management, inventory prediction, and routine customer communications.**
- Restaurant **teams provide genuine hospitality, c**ulinary creativity, problem-solving, and the personal touches that create memorable dining experiences.
- Customer relationships remain personal while service capabilities expand to meet modern expectations.
- Food quality stays high because **AI handles the logistics that used to distract from cooking and serving.**

Restaurant excellence is not about choosing between authentic hospitality and operational efficiency. **It is about using AI to make experienced restaurateurs more attentive, more consistent, and more profitable than ever before.**

Restaurants implementing AI create advantages that delivery apps and chain competitors cannot replicate. **The combination of instant responsiveness, personalized service, and genuine human warmth takes years to develop.** A chain restaurant is just a transaction. A family restaurant where the owner remembers your anniversary, where the server knows your kids' names, where the chef adjusts your regular order because he knows you are watching your sodium? That is hospitality no algorithm can fake.

Poppy's Pizza & Grill did not lose their family atmosphere when they implemented AI phone systems. They stopped missing calls during Friday dinner rush and started having real conversations with the customers who walked through their door. **The kitchen did not become impersonal when they added inventory AI.** They stopped running out of popular items on Saturday night and reduced waste by a third. The hosts did not become less welcoming when reservation management went digital. They stopped juggling phone calls and started actually greeting guests.

Here is what should keep restaurant owners who wait to implement AI up at night: While you are debating whether technology fits your family restaurant philosophy, the Italian place two blocks over just doubled their takeout capacity without adding staff. They answer every phone call during dinner rush. They remember customer preferences from six months ago. Their kitchen waste dropped 30% because AI predicted exactly how many steaks to prep.

The restaurant that implemented AI ordering six months ago never puts customers on hold. The pizzeria that added inventory prediction stopped running out of dough on their busiest nights. The café that automated their loyalty program knows exactly when regulars have not visited in a while and reaches out before they forget. **They are not selling out to technology. They are restaurant owners who decided that "please hold" was no longer acceptable when customers have a dozen other options on their phones.**

Your regulars have options. The delivery apps keep expanding. The chain restaurants keep getting more convenient. The ghost kitchens keep multiplying. **The relationships you have built over years matter, but relationships do not survive indefinitely when service quality falls behind what customers experience everywhere else.** The couple who celebrated every anniversary at your restaurant will still try the new place that remembers their dietary restrictions without being asked.

The labor market is not getting easier. The margins are not getting fatter. The customer expectations are not reverting to what they were before smartphones. The only question is whether you will use AI to become the restaurant that combines genuine hospitality with the responsiveness that modern diners demand, or whether you will keep telling yourself that good food alone will protect you from competitors who offer good food plus everything else.

Your recipes are not being automated. Your hospitality is not being outsourced. Your ability to make a stranger feel like family the moment they walk through your door is not something a chatbot can replicate. But the restaurants that wrap that warmth in AI-enhanced operations will fill more tables. They will turn more covers. They will build the neighborhood institutions that thrive while everyone else blames the economy for empty seats.

Here is what the chains understand that independent restaurants often lack: **technology does not replace hospitality. It creates space for it.** Every call your AI system answers is a server who can focus on the table in front of them. Every inventory prediction that prevents a stockout is a chef who can focus on cooking instead of emergency supply runs. Every automated reminder that brings back a lapsed customer is a relationship you would have lost to forgetfulness.

The restaurants that understand this are not becoming less personal. **They are becoming the kind of restaurant they always wanted to be, with AI handling the chaos so they can focus on the hospitality that made them fall in love with this business in the first place.**

How long before your regular customers start discovering that the restaurant down the street somehow always has time for them?

Chapter 20

Creative Industries

AI as Your Creative Collaborator

Creativity is not being replaced by AI—it is being supported to reach new heights of artistic and commercial success.

Creative industries face a unique AI challenge that goes beyond operational efficiency to the heart of artistic identity.

Graphic designers, photographers, writers, marketing agencies, video producers, and **other creative professionals** have built their careers on originality. Artistic vision and the capacity to address communication challenges through uniquely creative insight have traditionally been regarded as distinctively human attributes.

Yet AI tools for content generation, design automation, and creative workflow optimization are reshaping creative industries faster than any other sector. AI design tools like ***Midjourney***, ***DALL-E***, and ***Canva AI*** enable non-designers to create professional-looking graphics in minutes rather than days. Client expectations have evolved toward faster turnarounds, unlimited revisions, and lower costs that traditional creative processes cannot match economically. Content volume demands have skyrocketed as businesses require fresh, creative content for multiple social platforms, email campaigns, and digital touchpoints daily.

The dilemma becomes leveraging AI to meet modern production demands while preserving the originality and strategic thinking that justify creative expertise pricing.

Success Story
Santos Design

*Maria Santos** spent 12 years building her graphic design agency into a trusted partner for small and medium-sized companies. Her team of eight designers, writers, and strategists earned client loyalty through original thinking, cultural insight, and the ability to translate business objectives into compelling visual communications that generated measurable results.

But the creative landscape was transforming rapidly. AI design tools like ***Midjourney***, ***DALL-E***, and ***Canva AI*** were empowering individuals without a design background to produce high-quality graphics efficiently. Client expectations had evolved toward faster turnarounds, unlimited revisions, and lower costs that traditional creative processes could not match economically. Content volume demands had exploded as businesses needed fresh creative content for multiple social platforms, email campaigns, and digital touchpoints daily.

The creative dilemma: How do you use AI to meet-modern production demands while preserving the originality and strategic thinking that justify creative expertise pricing?

Trust: The Role of AI in Supporting Creative Work

Effective integration of AI in the creative industry is achieved when it enhances, rather than competes with, artistic vision and strategic goals. Positioning AI as an advanced tool that expands creative possibilities—rather than substituting for human ingenuity—is essential.

Scripts for Creative Partnership

For the senior designer concerned about artistic authenticity: "AI generates raw material and handles routine production tasks so you can focus on the conceptual thinking and creative direction that showcase your 12 years of design experience."

For the creative strategist worried about client relationships: "AI research and trend analysis give you deeper insights into audience behavior and competitive landscapes, making your strategic recommendations more powerful and data-driven."

For the junior creative concerned about learning: "AI handles production work so you can focus on developing design thinking and creative problem-solving skills. You will learn faster while working on more challenging creative concepts."

Building Creative Team Confidence in AI Partnership

Maria's implementation built enthusiasm across her creative team:

Creative Team Response:

- 94% of **designers indicated enthusiasm regarding AI creative tools.**
- Senior creative **professionals perceived AI as an opportunity to prioritize strategic and conceptual responsibilities.**
- Junior team members **valued the accelerated learning experiences provided by AI-assisted production.**
- Account managers noted **improved capabilities for client presentations.**

Client Response:

- 89% of respondents expressed a **positive response to enhanced creative capabilities** and accelerated iteration cycles.
- **Clients valued the transparency regarding increased efficiency** and broader opportunities for creative exploration.
- No issues were reported concerning the quality of creativity or originality of ideas.
- Multiple **clients inquired about additional creative services** made possible through AI adoption.

Ready: Preparing for Enhanced Creative Capabilities

Creative professionals have a complicated relationship with AI, and we understand why. You did not spend years developing your artistic eye and creative instincts just to watch a client type a sentence into ***Midjourney*** and declare themselves a designer. The anxiety is real. But here is the uncomfortable truth: **your clients are already using these tools, whether you like it or not.** The question is whether you will let AI make you irrelevant or use it to become indispensable.

Before you touch any AI creative tool, you need to get clear on what actually makes you valuable. If your business model depends on production volume and cranking out social media graphics or basic website layouts, it is important to know that AI is coming for that work, and no amount of preparation will save it. But if your value lies in creative strategy, brand thinking, and solving communication problems that require human judgment, **AI becomes your production department, not your replacement.** Be honest with yourself about which category you fall into.

The preparation that matters starts with your creative process, not the technology. Document how you actually work. When you take on a new client, what questions do you ask? How do you develop concepts? Where do you get stuck, and what eats up time that does not improve the final product? Most creative professionals have never mapped this out because the process lives in their heads. That needs to change. **AI tools can accelerate parts of your workflow, but only if you know which parts are worth accelerating and which ones require your full creative attention.**

Next, deal with the rights and ownership question before it becomes a problem. AI-generated content sits in a legal gray zone that is still being sorted out in courtrooms. If you use AI to create elements of client work, who owns what? Can your client trademark a logo that includes AI-generated components? What happens if an AI tool produces something that looks suspiciously like another artist's work? Your contracts need updating, and you need clear policies before you deliver your first AI-assisted project. **Winging it is not a strategy; it is a lawsuit waiting to happen.**

Then there is the client conversation. Some clients will love the idea that AI makes you faster and more affordable. Others will feel cheated if they find out a machine touched their project. You need to decide upfront how transparent you will be, and we would argue for full transparency. **Clients who discover you used AI without telling them will feel deceived, even if the work is excellent.** Clients who know from the start that AI handles production while you handle creative direction will respect the honesty and understand the value you bring.

Finally, pick your tools carefully. The creative AI landscape changes monthly, and chasing every new platform is a recipe for wasted time and half-learned software. Choose one or two tools that solve your specific bottlenecks, learn them deeply, and integrate them into your existing workflow before adding more. **A designer who masters one AI tool beats a designer who dabbles in six.**

The creative professionals who thrive will be the ones who use AI to do more of what makes them irreplaceable—not less.

Advance: Using AI in Creative Work

Maria's Agency Implementation

Maria's agency implementation and comprehensive creative transformation demonstrate how design agencies can achieve operational excellence while strengthening creative quality and client relationships.

Three Best AI Programs for Creative Industries

1. *Adobe Creative Suite with AI Features* - The Complete Creative Powerhouse

What it does: AI-driven design suggestions, automated video editing, intelligent font pairing, and content-aware tools integrated throughout the Adobe creative suite.

Why it is popular: Seamless AI integration enhances creative workflows without disrupting processes or formats.

Impact: Agencies report 60% less routine production time and 150% more creative iterations.

Cost: $52.99–79.49 per user/month for Creative Suite with AI.

Ideal for: Creative agencies needing advanced AI across multiple design fields.

2. *Canva Pro with AI Magic Tools* - The Accessible Creative Intelligence Engine

What it does: AI background removal, automatic design resizing for multiple platforms, smart design suggestions, brand kit integration, and collaborative creative workflows.

Why growing agencies adopt it: *Canva* offers professional AI capabilities with user-friendly interfaces, enabling both designers and non-designers to contribute to creative projects.

Real impact: Small creative agencies report 78% improvement in project turnaround times and 45% increase in client satisfaction through faster iteration and multi-platform optimization.

Monthly cost: $13-15 per user for Pro; Teams start at $15 a month for up to 5 users.

Best for: Small to medium creative agencies focused on social media, digital marketing, and multi-platform content creation.

3. Jasper AI for Creative Writing and Strategy - The Content Intelligence Amplifier

What it does: AI-powered copywriting assistance, brand voice consistency, content strategy development, campaign ideation, and creative brief enhancement.

Why strategic creatives choose it: ***Jasper*** combines content generation with strategic marketing intelligence designed specifically for creative professionals and agencies.

Real impact: Creative agencies report a 67% improvement in copy quality and an 89% reduction in initial draft time through AI-assisted writing and strategy development.

Monthly cost: $39-$69 per user for Creator and Pro plans, with Business plans available at custom pricing. The Creator plan suits individual content creators, while the Pro plan supports small marketing teams.

Best for: Creative agencies that need a comprehensive content strategy and copywriting support integrated with design capabilities.

90-Day Transformation for the Creative Discipline

Creative professionals face a unique challenge with AI: how do you use technology to enhance creativity without losing the authentic voice that clients hire you for? The answer is not the same for every creative discipline. A website designer needs different AI tools than a copywriter, and an advertising strategist requires different capabilities than a photographer.

This section organizes **AI implementation by creative discipline,** giving you specific tools and strategies for your particular craft.

Design & Visual Creative Services

Who This Covers: Graphic designers, website designers, UI/UX designers, brand consultants, logo designers

The Core Challenge: Clients hire you for your creative vision and aesthetic judgment, not to execute template designs. AI must speed up execution while preserving your distinctive creative voice.

Implementation Focus

AI handles repetitive design tasks, initial concepts, and technical execution. You provide creative direction, client understanding, and final aesthetic decisions.

Month 1: Foundation Setup

- Install and configure primary design AI tools.
- Build initial prompt libraries for your design style.
- Test AI on internal projects before client work.
- Document what AI handles well versus what needs human refinement.

Month 2: Client Integration

- Apply AI to 30% of client projects for time-intensive tasks.
- Use AI for concept exploration and rapid iteration.
- Maintain full human control of final client deliverables.
- Track time savings on specific project phases.

Month 3: Full Implementation

- AI supports 60-70% of initial design exploration.
- Human designers make all final aesthetic decisions.
- Client presentations emphasize your creative vision.
- Measure project turnaround improvements.

Three Best AI Tools for Design & Visual Services

1. Adobe Creative Suite with AI Features

What it does: Integrates AI capabilities directly into professional design tools you already use, including Photoshop's generative fill, Illustrator's vector generation, and automated design variations.

Why designers choose it: You do not need to learn entirely new software or change your workflow. The AI capabilities work within the professional tools that deliver client-ready files.

Real impact: *Santos Design Studio* in Portland reduced concept exploration time by 45%—distinctive brand aesthetic intact. Their 6-person team now handles 40% more client projects without adding designers.

Monthly cost: $60-85/month per designer for Creative Cloud with AI features

Best for: Established design firms with teams already trained on Adobe tools who want to add AI without disrupting existing workflows.

60-day milestone: Your team produces three concept variations in the time it previously took to create one, with clients commenting on faster turnaround without noticing quality changes.

2. *Webflow AI and 10Web AI Builder*

What it does: Creates functional website designs and generates responsive layouts from text descriptions or rough sketches. Handles technical coding while you focus on user experience and visual design.

Why designers choose it: The technology eliminates the technical barrier between your design vision and a working website. You design the experience; AI handles the code implementation.

Real impact: *Digital Agency* in Austin implemented ***Webflow AI*** and cut website development time from 6 weeks to 3 weeks for standard business sites. Their 4-person team now delivers custom websites at prices that compete with template-based competitors—design quality intact.

Monthly cost: *Webflow AI*: $29-49/month; ***10Web AI Builder***: $10-60/month

Best for: Website designers and digital agencies that want to focus on user experience and visual design rather than technical coding.

60-day milestone: You complete your first full custom website design in half the usual time, with the technical implementation matching your design vision without extensive developer involvement.

3. Figma AI and Design Assistant Plugins

What it does: Accelerates UI/UX design with intelligent layout suggestions, automated design system creation, and rapid prototyping capabilities for digital products and interfaces.

Why designers choose it: The tools integrate into collaborative design processes, helping teams maintain consistency while exploring more design directions in less time.

Real impact: *Product Design* in San Francisco uses ***Figma AI*** features to reduce design system documentation time by 60%. Their 5-person UX team now prototypes complete application flows in days rather than weeks.

Monthly cost: $15-45/month per designer

Best for: UI/UX designers, product design teams, and digital agencies working on application interfaces and design systems.

60-day milestone: Your design system documentation updates automatically as you work, and you produce interactive prototypes that feel like working products without writing code.

Writing & Content Services

Who This Covers: Copywriters, content writers, bloggers, SEO specialists, technical writers, grant writers

The Core Challenge: Clients pay you for your unique voice, subject matter expertise, and ability to connect with their specific audience. AI that makes your writing sound generic destroys your value.

Implementation Focus

AI handles research, first drafts of routine content, and optimization suggestions. You provide subject matter expertise, voice consistency, and the creative insights that make content compelling.

Month 1: Voice Development

- Train AI on your writing style and successful past work.
- Create detailed prompt libraries for different content types.
- Test AI on internal content before client projects.
- Establish editing workflow to maintain voice consistency.

Month 2: Strategic Integration

- Use AI for research and outline generation on all projects.
- Apply AI to first drafts of routine content types.
- Maintain full human control of strategic messaging.
- Track time savings on research and initial drafts.

Month 3: Full Workflow

- AI supports 50-60% of initial content creation.
- Human writers make all strategic decisions and final edits.
- Client work maintains your distinctive voice throughout.
- Measure productivity increases without quality loss.

Three Best AI Tools for Writing & Content Services

1. Jasper AI (formerly *Jarvis)*

What it does: Generates marketing copy, blog content, and creative writing using templates trained on high-performing content. Includes brand voice training and SEO optimization features.

Why writers choose it: The platform learns your writing style and maintains voice consistency across different content types. It understands marketing frameworks that writers use professionally.

Real impact: *Content Agency* in Denver uses ***Jasper AI*** for initial blog post drafts, reducing writing time by 35%—with the distinct voice their clients hired them for. Their 3-person team now handles 50% more clients without compromising quality.

Monthly cost: $49-125/month, depending on word count needs

Best for: Marketing copywriters, content agencies, and business bloggers who create regular content in established brand voices.

60-day milestone: You produce first drafts of routine content in one-third the usual time, with editing focused on strategic refinement rather than starting from scratch.

2. Copy.ai and Writesonic

What it does: Creates marketing copy, product descriptions, and social media content using AI trained on converting copy. Includes A/B testing suggestions and conversion-focused plans

Real impact: *Thompson Marketing Services* in Boston implemented ***Copy.ai*** and doubled their monthly output of product descriptions and ad copy without adding writers. Client conversion rates remained consistent while delivery speed improved significantly.

Monthly cost: ***Copy.ai***: $49/month; ***Writesonic***: $20-100/month

Best for: Copywriters focused on e-commerce, advertising, and direct response marketing who need volume without sacrificing conversion performance.

60-day milestone: You generate complete campaign copy variations in hours instead of days, testing more approaches to find what converts best for each client.

3. Surfer SEO with AI Content Generation

What it does: Combines SEO analysis with AI content generation, ensuring your writing ranks well, readability and expertise intact. Provides data-driven suggestions for content structure and keyword integration.

Why writers choose it: The technology handles the technical SEO requirements while you focus on creating valuable, readable content. You do not need to choose between writing for humans and ranking in search engines.

Real impact: *Digital Content* Company in Miami uses ***Surfer SEO*** to reduce content optimization time by 50% while improving client search rankings. Their 4-person team maintains editorial quality while meeting aggressive SEO targets.

Monthly cost: $89-219/month depending on content volume

Best for: Content writers and SEO specialists who need to balance search ranking performance with authentic, valuable content.

60-day milestone: Your content ranks higher in search results without the robotic keyword stuffing that makes readers leave, and you spend less time on technical optimization.

Advertising & Campaign Management

Who This Covers: Ad agencies, media buyers, campaign managers, PPC specialists, and creative directors managing campaigns.

The Core Challenge: Advertising requires both creative excellence and data-driven optimization. AI must enhance both without replacing the strategic thinking that delivers results.

Implementation Focus

AI handles ad variation generation, budget optimization, and performance testing. You provide strategic positioning, creative direction, and the client understanding that drives campaign success.

Month 1: Testing & Learning

- Implement AI tools on existing campaigns for comparison.
- Test AI-generated ad variations against human-created versions.
- Document performance patterns and best use cases.
- Build prompt libraries for your most successful ad styles.

Month 2: Strategic Scaling

- Use AI for 50% of ad variation generation.
- Apply AI to budget allocation and bid optimization.

- Maintain human control of campaign strategy and messaging.
- Track performance improvements and time savings.

Month 3: Full Integration

- AI generates 70-80% of ad variations for testing.
- Human strategists make all positioning and budget decisions.
- Client reporting emphasizes strategic results.
- Measure campaign performance and efficiency gains.

Three Best AI Tools for Advertising & Campaign Services

1. *AdCreative.ai* Generative AI Marketing

What it does: Generates complete ad creatives, including images, headlines, and body copy, optimized for different platforms. Uses conversion data to improve recommendations over time.

Why agencies choose it: The technology produces platform-specific variations that look professionally designed, not template-based. You focus on strategy while AI handles execution.

Monthly cost: $29-149/month, depending on generation volume

Best for: Digital advertising agencies and media buyers who need high-volume ad testing and creative variation.

60-day milestone: Your campaigns test significantly more creative variations, with data showing which approaches perform best before you commit significant ad spend.

2. *Creatopy AI* (formerly *Bannersnack*)

What it does: Creates animated display ads, social media graphics, and video ads using AI-powered design and animation tools. Includes brand consistency features and multi-platform resizing.

Monthly cost: $49-149/month, depending on features and user count

Best for: Advertising agencies creating display advertising and social media campaigns across multiple platforms.

60-day milestone: You launch campaigns across multiple platforms with perfectly sized creative variations, completing in days what previously required weeks.

3. *Google Ads Performance Max with AI Optimization*

What it does: Uses machine learning to automatically optimize ad delivery, bidding, and placement across Google's advertising network. Combines search, display, video, and shopping into unified campaigns.

Why agencies choose it: The technology finds conversion opportunities across channels that human analysis might miss, while you control budget, strategy, and creative messaging.

Monthly cost: No additional cost beyond Google Ads spend (percentage fee varies by agency)

Best for: PPC specialists and agencies managing comprehensive Google Ads campaigns who want automated optimization within strategic guardrails.

60-day milestone: Your campaigns automatically adjust bidding and placement based on conversion data, delivering better results without constant manual optimization.

Media Production Services

Who This Covers: Video producers, photographers, podcasters, audio producers, videographers, content creators

The Core Challenge: Media production combines technical execution with creative vision. AI must streamline production workflow without compromising the creative quality that clients value.

Implementation Focus

AI handles time-intensive editing tasks, initial cuts, and technical corrections. You provide creative direction, storytelling expertise, and the production quality that defines your brand.

Month 1: Production Testing

- Implement AI on post-production workflow for completed projects.
- Test AI editing capabilities on archive footage.
- Establish quality standards that AI must meet.
- Document time savings on routine tasks.

Month 2: Workflow Integration

- Use AI for initial video cuts and photo selections.
- Apply AI to color correction and audio cleanup.
- Maintain human control of final creative decisions.
- Track production efficiency improvements.

Month 3: Full Production Enhancement

- AI handles 60-70% of technical post-production work.
- Human producers make all creative and strategic decisions.
- Client deliverables showcase your creative vision.
- Measure turnaround improvements and capacity increases.

Three Best AI Tools for Media Production Services

1. Descript

What it does: Enables text-based video and audio editing, automatic transcription, AI voice cloning for corrections, and multi-track editing without traditional timeline complexity.

Why producers choose it: You edit video by editing text, dramatically reducing the technical skill barrier for clients who want to make changes. The AI handles audio cleanup and filler word removal automatically.

Monthly cost: $24-50/month per editor

Best for: Podcast producers, video editors, and content creators who produce interview-based or narrative content requiring significant editing.

60-day milestone: Your post-production workflow transforms from timeline-based editing to text editing, with clients able to request changes by simply marking up a transcript.

2. Adobe Premiere Pro with AI Features (Sensei)

What it does: Integrates AI-powered color matching, scene detection, auto-reframing for different aspect ratios, and speech-to-text all within professional video editing software.

Why producers choose it: The AI capabilities work within the professional tools you already know, adding intelligence without forcing workflow changes. You maintain full creative control while AI handles technical optimization.

Monthly cost: $23-85/month, depending on individual or team plans

Best for: Professional video production companies that need AI capabilities integrated into comprehensive editing workflows.

60-day milestone: Your team produces color-matched multi-camera edits and platform-specific aspect ratios automatically, focusing creative time on storytelling rather than technical consistency.

3. Runway ML for Video Effects and Generation

What it does: Provides AI-powered video effects including background removal, object tracking, motion smoothing, and even generative video from text descriptions for B-roll and effects.

Monthly cost: $12-76/month depending on generation volume and features

Best for: Video producers and creative agencies that need advanced effects and creative transitions without extensive motion graphics expertise.

60-day milestone: You add professional background replacements and motion effects to client videos in minutes rather than hours, expanding your creative capabilities without hiring specialists.

Implementation Success Across All Creative Disciplines

Regardless of your creative specialty, successful AI implementation follows the same pattern: AI amplifies your creative capabilities without replacing your distinctive voice and expertise. Clients continue hiring you for your creative judgment, strategic thinking, and unique perspective—they just get faster turnaround and more exploration of creative directions.

The competitive advantage comes from using AI to do more high-quality creative work in less time, not from replacing human creativity with automated templates. Your business grows because you deliver better results faster, not because you cut corners.

Connect: Making AI and Creative Vision Work Together

Everything clicks for creative industries when AI analytical power combines with human artistic vision to create communication solutions that neither could achieve alone.

The Complete Creative Integration Pattern

AI implementations in the creative industry commonly adhere to a similar integration pattern:

AI Handles: Routine production tasks, design variations, content adaptation, research compilation, and technical optimization.

Human Creativity Provides: Expertise includes conceptual development, strategic planning, artistic direction, cultural analysis, and client relationship management.

Working Together Results: Enhanced opportunities for creative exploration, accelerated iteration processes, improved client solutions, and the preservation of artistic integrity.

Real-World Creative Integration Success

Brand Strategy Development:

AI Research: Conducted an in-depth analysis of target audience behavior, competitor creative strategies, and prevailing industry design trends.

Creative Strategy: Senior Strategist contextualized AI-driven insights with client business goals and brand identity, formulating a distinctive positioning strategy.

Superior Outcome: Delivered a data-informed creative approach that resulted in a 234% increase in campaign performance while preserving the brand's unique voice.

Design Production Enhancement:

AI Production: Automated tools create multiple design versions and optimize campaigns for various platforms.

Creative Direction: The art director shapes concepts and upholds quality standards to drive strong creative outcomes.

Enhanced Result: Delivering 150% more creative options without sacrificing quality or client satisfaction.

Maria's journey shows how AI enhances every part of agency operations:

> **Strategic Planning:** Maria used AI research to spot trends and refine strategies.
>
> **Creative Development:** Her team quickly explored concepts with AI, ensuring brand authenticity.
>
> **Client Presentation:** AI-generated options impressed clients and highlighted strategic depth.
>
> **Production:** Automated adaptations saved time and maintained brand consistency.
>
> **Account Management:** AI-driven insights improved campaign performance and strengthened client relationships.

Keep: Maintaining Creative Excellence

Creative businesses that successfully adopt AI build advantages that grow over time by developing systems that boost creative skills—without losing the artistic vision that defines their value.

Creative Industry Leadership

Maria's Agency Recognition:

- **Featured in design publications** as a model for combining AI efficiency with creative excellence.
- **Speaking opportunities at creative conferences** about AI integration in design workflows.
- **Founded to demonstrate AI** as a means to enhance, not replace, creative thinking.

Artistic Innovation Through Technology

- **Present AI adoption as a commitment to innovation** and superior client service, not just cost savings.
- **Use AI capabilities to expand creative possibilities and strategic insights** that are not available through traditional methods.
- Show how **technology enhances artistic vision** and creative problem-solving.
- **Build a reputation as a creative leader** who combines innovative capabilities with authentic artistic integrity.

The Creative Industry Advantage

The most successful creative professionals do not treat AI as a threat to their artistic vision or a replacement for the creative instincts they spent years developing. They use **AI as a production amplifier** that requires experienced creatives to fully realize its potential.

Key Success Elements:

- AI handles routine production tasks, initial concept variations, and technical optimization.
- Creatives provide strategic thinking, artistic direction, cultural insight, and client relationship management.
- Client relationships remain personal while creative capabilities expand beyond what human-only production can deliver.
- Creative integrity stays intact because humans make every strategic and artistic decision.

Creative excellence is not about choosing between artistic vision and production efficiency. **It is about using AI to make experienced creatives more prolific, more exploratory, and more valuable than ever before.**

Creative businesses thoughtfully using AI create advantages that template platforms and overseas production farms cannot replicate. **The combination of rapid iteration, strategic insight, and distinctive artistic voice takes years to develop.** A Canva template is just a starting point. A designer who can interpret a client's brand strategy, understand their audience at a gut level, and create work that makes people feel something specific? That is creative expertise no prompt can generate.

The photographer who implemented AI editing workflows did not lose her artistic eye. She stopped spending nights color-correcting routine shots and started taking on projects that actually challenged her creatively. The content writer who added AI drafting tools did not sacrifice his voice. He eliminated the blank-page paralysis and spent his energy on the strategic messaging and turns of phrase that clients actually pay premium rates for. The video production company that automated rough cuts did not become a content mill. They doubled their client capacity while maintaining the storytelling quality that built their reputation.

Here is what should keep creative professionals who wait to implement AI up at night: While you are deciding whether AI compromises your artistic integrity, your competitor down the street just delivered three concept directions in the time it took you to finish one. You are still billing for production hours. They are billing for creative strategy.

The designer who implemented AI production tools six months ago offers more options, iterates faster, and spends client meetings discussing brand positioning rather than explaining why revisions take so long. The copywriter who automated research and first drafts delivers campaigns in days that used to take weeks. The photographer who added AI culling and editing handles twice the weddings without the Sunday night editing marathons. **They are not selling out. They are creatives who decided that "good work takes time" was no longer a competitive advantage when clients can get good work that takes less time.**

Your clients have options. The template platforms keep improving. The AI image generators get more sophisticated every month. The overseas production teams offer prices you cannot match on volume alone. **The relationships you have built over years matter, but relationships do not survive indefinitely when turnaround times and creative output fall behind what clients can get elsewhere.** The marketing director who loves your work will still take a meeting with the agency that delivered a full campaign while you were still scheduling the kickoff call.

The creative landscape is not reverting to what it was before AI. The client's expectations are not improving. The competitors who figured this out first are not slowing down to let you catch up. **The only question is whether you will use AI to become the creative who combines distinctive artistic vision** with the speed and exploration that modern clients demand, or whether you will keep telling yourself that craft alone will protect you from creatives who offer skill plus everything else.

Your artistic eye is not being automated. Your creative instincts are not being outsourced. Your ability to understand what a client actually needs and translate that into work that moves people is not something a prompt can replicate. But the creatives who wrap that vision in AI-enhanced production will win the projects. They will command the creative fees. They will build the studios and agencies that thrive while everyone else blames AI for stealing their work.

Here is the truth about the creative industry that surprises everyone: **AI is not coming for your job. AI is coming for the parts of your job you never loved anyway.** Boring file organization, routine edits, or the hundredth variation on the same concept. The production grind that kept you from the creative thinking that made you fall in love with your work in the first place. The creatives who understand this are not mourning the death of their craft. **They are rediscovering why they became creatives in the first place, with AI handling the tedium so they can focus on the work that actually matters.**

Have you considered when your most valued clients might begin to question why another creative partner consistently delivers more ideas and quicker results?

Chapter 21

Transportation & Logistics

Smarter Services with AI Systems

Your route planning and customer service expertise can be more valuable when **AI handles route planning and schedules.** This allows you to focus on complex logistics and client relationships.

Transportation and logistics businesses face unique AI implementation challenges because success relies on real-time problem-solving, safety management, and customer service. They must adapt to unpredictable conditions like weather, traffic, equipment failures, and last-minute customer changes. Delivery companies, freight brokers, moving services, and transportation providers have gained competitive advantages through local knowledge, relationship management, and operational expertise that seemed impossible to replicate with technology.

This chapter demonstrates how **transportation professionals can utilize AI to enhance logistics** expertise, creating competitive advantages that leverage industry knowledge while delivering unprecedented improvements in efficiency and customer service.

Success Story
Topper's Transportation

Over 14 years, *Mike Topper** built his regional trucking company, serving manufacturers throughout Texas and Oklahoma. His eighteen drivers and twelve trucks moved everything from auto parts to construction materials, competing with national carriers through personalized service and regional expertise that kept customers loyal, even when larger competitors offered lower rates.

However, **the logistics landscape was evolving rapidly,** and Mike was losing contracts to competitors who could provide instant quotes, real-time tracking, and predictive delivery windows that his manual systems could not consistently match.

The essential consideration for every transportation owner: How can you deliver the efficiency and convenience customers now demand, while preserving the personalized service and flexibility that have established your reputation?

Mike's solution was not to compete with technology alone. **He used AI to enhance his regional knowledge and customer relationships while delivering the digital experience that customers demanded.**

Trust: Getting Buy-In from Drivers and Customers

Transportation AI implementation faces unique challenges because **both drivers and customers directly experience the impact of technological changes.** Additionally, both groups express concerns regarding job and economic security, as well as the quality of services provided. In this industry, cultivating personal relationships is integral to building and maintaining customer loyalty.

Scripts for Driver Conversations

Transportation success depends on experienced drivers who understand routes, customers, and equipment. The trust discussion involves presenting AI in a way that emphasizes its contribution to enhancing the driver's experience, rather than replacing it.

For the experienced driver concerned about job security: "This route optimization system handles the paperwork and coordination so you can focus on safe driving and maintaining customer relationships that make our company different from the big carriers."

For the driver worried about technology complexity: "The system sends optimized routes directly to your phone using the same GPS you already know how to use. You still make all the driving decisions based on your experience with traffic, weather, and customer needs."

For the safety-conscious driver concerned about monitoring: "The vehicle monitoring helps you stay safe and proves to customers and insurance companies that you are the professional driver they can trust. It protects your reputation and our company's safe standing."

Ready: Preparing for Enhanced Logistics Operations

Transportation is an unforgiving business. A delayed shipment does not care that your routing software crashed. A customer whose freight sits on a dock for three days does not want to hear about your AI implementation timeline. **Before you add any technology to your operation, you need to be certain your foundation can handle change because in this industry, the cost of getting it wrong shows up fast.**

There is no shortage of trucking company owners who bought expensive fleet management systems only to discover their drivers refused to use them. **The technology worked fine; the implementation failed because nobody asked the people behind the wheel what they actually needed.**

Your drivers have years of route knowledge, customer relationships, and problem-solving instincts that no algorithm can replace. If they see AI as a surveillance system or a threat to their jobs, they will find ways to work around it. **Getting their buy-in is not a soft skill; it is a business requirement.**

Start your preparation by auditing what you actually know about your operation. How accurate are your delivery time estimates? Not what you tell customers, but what actually happens. Pull your data from the last six months and calculate your on-time percentage honestly. **If you are running at 85% and telling customers 95%, AI is going to expose that gap, not fix it.** You need to understand your real performance before you can improve it.

Next, look at your existing technology stack. Can your current dispatch system talk to a route optimization tool? Does your accounting software integrate with fleet management platforms, or will someone be manually entering data twice? **Transportation AI tools are only as good as the information flowing into them. If your systems do not connect, you are building on sand.** Sometimes, the right preparation is not buying new AI; it is upgrading your basic infrastructure.

Then get serious about your data quality. AI route optimization needs accurate vehicle capacity, real delivery windows, and current customer addresses. AI predictive maintenance needs consistent service records and mileage tracking. **If your maintenance logs are incomplete or your customer database has not been cleaned since 2019, that is your starting point.** We have watched transportation companies spend months "training" AI systems when the real problem was garbage data from day one.

Finally, know your numbers cold. What does a late delivery actually cost you, in penalty fees, lost future business, and reputation damage? What is your fuel cost per mile, and how much variability exists between drivers on the same routes? **What percentage of your revenue comes from customers who chose you specifically because of personal relationships, versus customers who would switch for a 5% discount?** These answers tell you where AI can create real value and where it is just an expensive experiment.

The transportation companies that succeed with AI treat preparation the way they treat pre-trip inspections: thoroughly documented, and non-negotiable.

Success Story

Glo Pals

Glo Pals, is a sensory toy company that originated as a project at Mississippi State University. It shows how **small businesses can use AI-enhanced logistics to compete** with larger companies—without losing personalized customer service.

The Challenge: *Glo Pals* produces liquid-activated educational toys for kids that light up when used. As Glo Pals grew from a modest team of two to a workforce of 27, and expanded operations to include international shipping, the company encountered significant logistical challenges common to small businesses. These included spending excessive time negotiating with freight forwarders, experiencing lengthy delays while waiting for shipping quotes, and struggling to efficiently manage multiple shipments at once. These hurdles underscored the need for a more streamlined approach to logistics management as the company scaled its operations.

The Solution: *Glo Pals* implemented ***Freightos*** marketplace platform to access instant freight quotes and streamline booking across multiple logistics providers.

Logistics Implementation Strategy:

- Freight **quotes are available in real time** from several providers, replacing manual negotiation processes.
- Instant **booking capabilities eliminated waiting time** for coordinating transportation.
- Integrated shipment tracking from origin to destination warehouse.
- Comprehensive 24/7 customer support is available to assist with all transport-related issues.

Results after 18 months:

- **Shipping costs decreased by almost 20%** through competitive quote comparisons.
- **Quote turnaround now takes seconds** instead of hours or days.
- Streamlined shipment management through a centralized dashboard.

- Enabled rapid scaling to support partnerships with Sesame Street, Kohl's, SeaWorld, Barnes & Noble.

Glo Pals founder Hagan's insight: "We're big fans of ***Freightos.com***. I love being able to log in and see all current shipments and approximate delivery dates. It allows us to compare quotes on demand and it's so nice not to have to wait hours or days to get quotes back from a freight forwarder, just to find out they are more expensive than what is instantly available on ***Freightos.com***."

Advance: Putting AI to Work in Transportation

Regional Trucking Implementation

Mike's 90-day transformation showcases how regional transportation companies can achieve significant improvements while preserving the local expertise and customer relationships that distinguish them from national carriers.

30-Day Foundation Period:

Route4Me AI Optimization offers basic route planning and customer communication.

- Efficient multi-stop route planning **cut fuel costs by 23%.**
- Real-time tracking eliminated 78% of "where is my delivery" customer calls.

- **Driver productivity increased 19%** through better route coordination and reduced deadhead miles.
- Customer satisfaction improved through proactive delivery updates and accurate ETAs.

60-Day Capability Expansion:

Samsara Fleet Management integration for comprehensive vehicle monitoring and driver coaching.

- Predictive maintenance alerts identified three instances of potential breakdowns prior to any impact on deliveries.
- Driver safety coaching resulted in **a 34% reduction in hard braking incidents and contributed to enhanced fuel efficiency.**
- **Optimizing vehicle utilization led to a 28% increase in capacity** without requiring additional trucks.
- **Insurance costs decreased through documented safety improvements** and reduced incident rates.

90-Day Optimization Period:

- Combining both platforms **led to logistics intelligence** far beyond competitors' capabilities.

- **Customized delivery algorithms** have been developed to address specific regional routes and cater to customer preferences.
- **Automated customer communication** delivered service standards comparable to those of national carriers.
- Driver analytics enabled coaching that boosted safety and efficiency.
- **Market position strengthened** as a technology-driven regional transportation specialist.

Complete Results after 6 months:

- Achieved **67% revenue growth** by maximizing capacity and delivering competitive logistics services.
- **Customer retention reached 94%,** attributed to improvements in reliability and customer communication.
- **Operating margins increased 31%** through fuel savings and operational efficiency.
- **Driver satisfaction improved** as technology eliminated frustrating coordination delays.
- The company's competitive position has been improved, establishing the company as a regional carrier equipped with national-level capabilities.

Three Best AI Programs for Transportation and Logistics

1. Route4Me with AI Optimization - The Complete Delivery Intelligence Engine

What it does: AI-powered multi-stop route optimization, real-time traffic integration, customer communication automation, driver performance analytics, and delivery confirmation tracking designed specifically for small to medium transportation firms.

Why transportation professionals choose it: Route4Me serves tens of thousands of businesses and is among the top five highest-grossing navigation apps worldwide. Founded in 2009, it has helped more than 1.9 million users efficiently plan and optimize multiple-stop driving routes.

Real impact: ***Route4Me*** combines Google Maps accuracy with patented Dynamic Route Optimization algorithms to reduce empty miles and improve customer service for businesses by managing dozens, hundreds, or thousands of daily stops.

Monthly cost: $39-199 per vehicle, depending on features and the fleet size.

Best for: This solution is designed for small to medium-sized delivery firms, field service organizations, and businesses with numerous daily destinations that require robust route planning capabilities yet wish to avoid the intricacies associated with enterprise-level systems.

2. Samsara Fleet Management with AI Analytics - The Safety and Efficiency Amplifier

What it does: AI-powered vehicle monitoring, predictive maintenance alerts, driver safety analysis, fuel efficiency optimization, ELD compliance, and comprehensive fleet analytics that are designed for growing businesses.

Why fleet managers adopt it: ***Samsara*** specializes in working with transportation and logistics, construction, food and beverage, government, and field services companies, providing near-enterprise-level monitoring that is easy enough for small teams to master quickly.

Real impact: Univar Solutions reduced harsh driving events by 40% using Samsara's AI dashcams. Transportation companies report significant improvements in safety incidents and fuel efficiency through AI-enhanced monitoring and driver coaching.

Monthly cost: $40-80 per vehicle, depending on features and monitoring capabilities.

Intended for: Small to medium fleets with 5-50 vehicles that require safety management, driver development, and operational efficiency features, and seek enterprise-level capabilities without added complexity.

3. Freightos with AI Freight Matching - The Logistics Relationship Engine

What it does: An AI-powered freight marketplace providing instant rate comparisons, automated booking, shipment tracking, and carrier network management designed for small to medium freight users.

Why businesses choose it: ***Freightos*** gives businesses without logistics experience access to competitive freight rates and bulk-buying power that were once reserved for large companies.

Real impact: Businesses like Glo Pals achieve a 20% reduction in shipping costs and eliminate quote wait times, reducing delays from days to hours and seconds. Over 12,000 businesses use the **Freightos** platform to instantly access 45+ logistics providers.

Monthly cost: Free to use the marketplace with transparent booking fees.

Best for: Small to medium importers, exporters, and e-commerce businesses who need professional freight services without maintaining relationships with multiple logistics providers.

Connect: Making AI and Logistics Expertise Work Together

The shift occurs for transportation businesses when **AI analytical power partners with human logistics expertise** to create service capabilities that neither could achieve alone.

By integrating advanced AI technologies with the nuanced understanding and experience of logistics professionals, **companies can unlock operational efficiencies, improve safety outcomes, and offer**

enhanced customer service. AI excels at processing vast amounts of data, identifying patterns, and automating repetitive tasks such as route optimization, real-time traffic analysis, and load matching. However, human expertise remains essential for complex problem-solving, managing customer relationships, making safety judgments, and adapting strategies based on local or situational knowledge.

The Power of Human-AI Collaboration in Transportation

When technological strengths are combined, transportation businesses gain access to tools and insights that drive more informed decisions, reduce costs, and foster sustainable growth. This synergy boosts individual skills and enhances service potential, enabling companies to adapt quickly to industry changes.

The Complete Transportation Integration Pattern

Every successful transportation implementation follows the same integration pattern:

AI Handles: Route optimization, real-time traffic analysis, fuel efficiency monitoring, maintenance scheduling, customer communication automation, and load matching.

Human Expertise Provides: Complex problem-solving, customer relationship management, safety judgment, and crisis response while utilizing local knowledge to oversee strategic route planning.

Working Together Results: Greater efficiency, better service, lower costs, and stronger competitiveness.

Real-World Integration Success

Load Matching Enhancement:

AI Analysis: ***Freightos*** system analyzed carrier capacity, geographic patterns, and market conditions to identify optimal freight matches for Mike's regional routes.

Human Expertise: Mike leveraged 14 years of regional knowledge to evaluate carrier reliability, customer requirements, and seasonal patterns that AI data might not fully capture.

Enhanced Outcome: 67% improvement in load matching efficiency with maintained carrier relationships and customer satisfaction through AI-enhanced decision-making guided by experienced judgment.

Route Optimization Integration:

AI Coordination: ***Route4Me*** analyzed traffic patterns, delivery windows, and vehicle capacity to optimize daily routes across Mike's 18-driver fleet.

Logistics Expertise: Drivers and dispatchers used local knowledge of construction zones, customer preferences, and regional challenges to refine AI recommendations.

Superior Results: A 35% reduction in fuel costs has been achieved by implementing technology-driven route optimization, complemented by regional expertise and strong customer relationships, resulting in enhanced customer service.

Keep: Maintaining Transportation Excellence

Transportation businesses that successfully implement AI create advantages that strengthen over time. By establishing systems that enhance logistics capabilities, preserving industry knowledge, and strengthening customer relationships that define their value.

Market Position Through Innovation

- Frame AI adoption as an initiative to enhance driver safety and customer service, rather than solely focusing on cost savings.
- Utilize technological capabilities to broaden service offerings and improve reliability for regional customers.
- Demonstrate how AI adds value—without diminishing the local expertise and connections that support current pricing structures.

The Transportation and Logistics Advantage

The most successful transportation companies do not treat AI as a surveillance system or a replacement for experienced drivers. They use it as an operations tool that requires skilled logistics professionals to fully realize its potential.

Key Success Elements:

- **AI handles route optimization,** real-time tracking, and predictive maintenance alerts.
- **Drivers and dispatchers provide local knowledge,** customer relationships, and crisis judgment.
- **Customer relationships remain personal** while service capabilities expand.
- **Safety culture strengthens** as technology and experience work together.

Transportation excellence is not about choosing between regional expertise and digital efficiency. **It is about using AI to make experienced drivers and dispatchers more capable, valuable, and competitive than ever before.**

Small and regional transportation companies thoughtfully using AI create advantages that national carriers cannot easily reproduce. **The combination of advanced logistics technology, personal service, and local knowledge takes years to develop.** A truck with optimized routes is just efficient. A truck with optimized routes, a driver who knows that the loading dock at Henderson Manufacturing is always backed up on Tuesdays, and a dispatcher who calls customers before problems become complaints? That is a business relationship no algorithm can replicate.

Here is what should keep transportation owners who wait to implement AI up at night: While you are deciding whether route optimization is worth the monthly fee, your competitor across town just cut their fuel costs by 23% and started offering real-time tracking that makes your "call us for updates" approach look like a relic. Their drivers are finishing routes faster with less frustration. Their customers are getting delivery confirmations before they even think to ask.

The freight broker who implemented AI load matching six months ago is now handling twice the volume with the same staff. The regional carrier that added predictive maintenance caught three potential breakdowns before they stranded drivers on the highway. **They are not technology companies. They are transportation professionals who decided that good enough was no longer good enough.**

Your customers have options. Your drivers have options. Your business reputation, built over years of reliable service, can erode faster than you think when a competitor offers the same personal attention plus the digital experience that customers now expect from everyone.

How long before your best customers start taking calls from your competition?

Chapter 22

Real Estate & Property Management

Your market knowledge and client relationships are not becoming obsolete—they are becoming more powerful through **intelligent systems that handle routine coordination while you focus on strategic guidance and relationship building** that create successful transactions.

Real estate professionals face unique AI implementation challenges because success depends entirely on relationship management, market expertise, and the ability to guide clients through complex emotional and financial decisions during major life transitions. Buyers and sellers choose agents for local knowledge, negotiation skills, and trusted guidance that technology cannot replicate, yet they increasingly expect the convenience and communication that AI-enhanced platforms provide.

This chapter reveals how **real estate professionals can use AI** to enhance rather than replace expertise and relationships, creating competitive advantages that honor industry knowledge while delivering unprecedented client service and market insight.

Success Story
The DeBerry Team

Roxanne DeBerry has been selling homes in Plano, Texas since 1975. That is not a typo. For nearly fifty years, **she has built her reputation one handshake at a time, one kitchen table conversation at a time, one referral at a time.** When her son Larry joined the family business as marketing manager, he brought fresh eyes to an old problem that had been nagging the team for years.

"We knew the expectation," Larry says. "Get back to online leads in under two minutes. We were nowhere near that."

The *DeBerry Team* runs super lean. Two buyer agents, one seller specialist, and four support staff. That is the whole operation. When leads came flooding in from Google, Facebook, and their website at eleven o'clock at night, no one was answering. The family was asleep. **The leads were growing cold.**

They considered hiring an inside sales agent to handle lead qualification, but the math did not work. A full-time ISA costs around sixty thousand dollars annually in salary alone, not counting benefits, training, and management overhead. For a seven-person team already stretched thin, that expense would eat into margins without any guarantee of results.

Larry started searching for alternatives and found ***Structurely***, an AI assistant platform designed specifically for real estate lead qualification. The system works like having a tireless inside sales agent who never sleeps, never takes lunch breaks, and never calls in sick.

How It Works

A prospective buyer finds the *DeBerry Team* website at 11:47 PM and submits an inquiry. Within seconds, the **AI assistant begins a natural-language conversation, asking about their timeline, budget, and which neighborhoods interest them.** The exchange feels personal enough that leads often have no idea they are talking to artificial intelligence.

Larry describes watching one of these conversations unfold: "I was in bed at midnight watching this lead chat back and forth with our assistant for fifteen minutes. It was really cool because they thought they were talking to a regular person."

The AI does not try to close deals. **It qualifies leads, gathering the essential information agents need to have productive conversations.** When a lead demonstrates genuine buying intent, the system alerts the appropriate agent and hands off the relationship. Cold or unresponsive leads stay in automated nurture sequences until they warm up.

The Results

Over sixteen months, the *DeBerry Team's* AI assistant processed more than **1,200 leads** and **sent 15,000 messages**. More than half of those leads came in after business hours, meaning they would have received no response at all under the old system.

The engagement rate hit **40 percent**, roughly four times what most real estate teams achieve with traditional follow-up methods. The assistant handed off 101 qualified leads directly to agents, each one already vetted with timeline, budget, and property preferences documented.

Most importantly, the *DeBerry Team* kept doing what they do best. Roxanne still meets with sellers face-to-face. The buyer agents still walk clients through homes. **Nobody is stuck making cold calls at nine in the morning when they should be showing houses.**

"We have full trust in the relationship with the assistant to deliver buyer leads," Larry says. "Some of our agents like to jump in early and take over conversations. Others prefer to wait until the lead is fully qualified. Both approaches work."

The family business Roxanne built in 1975 now operates with technology that would have seemed like science fiction back then. But the fundamentals have not changed. Relationships still drive everything. AI just makes sure no opportunity slips through the cracks while the team is busy serving the clients already in their care.

Trust: Building Confidence in AI-Enhanced Real Estate Service

Real estate agents panic when AI tools start suggesting listing prices or predicting market trends. They either blindly trust the algorithm because it looks sophisticated, or they ignore it completely because they do not trust a computer with their clients' biggest financial decisions. Both reactions miss the point. AI market analysis tools are not crystal balls, and they are not trying to replace your judgment—they are data aggregation systems that can process comparable sales and market patterns faster than you can pull records manually.

The real question is not **whether you should trust AI with pricing decisions. The question is whether you understand what the system is analyzing and why it matters.** If you cannot explain an AI valuation to your client in terms of actual neighborhood trends and recent sales, you have no business presenting it as professional advice. Let me show you how to build genuine confidence in these tools without becoming either an algorithm follower or a technology skeptic.

Client Trust Plans

Real estate clients hire you for judgment and guidance that they consider beyond the capabilities of AI. The trust conversation requires **positioning AI as enhancing your expertise,** making it more comprehensive and responsive, rather than replacing it.

Script for Addressing Client AI Questions

For clients concerned about AI determining their home value: "AI helps me analyze market data and identify pricing patterns faster than ever before, but you are still getting my experience, my negotiation strategy, and my personal guidance on positioning your specific property in this neighborhood."

For clients worried about data security: "All AI tools we use are specifically designed for the real estate industry with the highest security standards. Your information remains completely protected, and AI just helps me serve you more efficiently throughout the transaction."

For clients questioning the value of agent representation: "AI handles the routine data gathering and market analysis that used to take hours, so I can spend more time on strategic pricing, negotiation, and providing the personalized guidance you need during this major life decision."

Real Estate Trust Framework

Every successful real estate AI implementation requires three trust conversations:

With Experienced Agents: Position AI as a career advancement opportunity that makes veteran market knowledge more valuable through enhanced analytical capabilities and superior client service.

With New Agents: Frame AI proficiency as an essential modern professional skill that accelerates career development and increases value to clients and brokers.

With Clients: Emphasize AI as a tool for more comprehensive market analysis and faster communication—plus the personal attention and expert guidance through complex transactions.

Real-World Trust Building Results

Larry's implementation exceeded the expectations of his team and clients.

Team Response:

- 91% of **agents expressed confidence in AI real estate integration plans.**
- New agents showed enthusiasm for AI training and modern technology adoption.
- **Veteran agents acknowledged AI as a necessary competitive edge** in an evolving market.
- Administrative **staff valued the decreased burden of routine data entry** and scheduling.

Client Response:

- 88% **positive reaction to service enhancement communication.**
- Clients appreciated transparency about improved market analysis capabilities.
- No concerns raised regarding data security or service quality.
- Several **clients requested information about expanded property management services.**

This pattern holds true across all types of real estate businesses. When you position AI as enhancing, over replacing expertise, both team members and clients respond positively.

Ready: Preparing Systems and People

Real estate agents love to talk about relationships, but when asked to produce their client database, we often get a shoebox full of business cards and a Gmail account with 4,000 unorganized emails. **You cannot build AI-enhanced client service on top of chaos.** Before you invest in any market analysis tool or automated communication platform, you need to know exactly what you are working with and be honest about its condition.

Here is the uncomfortable question: how many past clients could you contact right now with accurate information? Not "I think I have their number somewhere," but actually reach them with a personalized message about market conditions in their neighborhood? **If that number is less than 80% of everyone you have ever closed a deal with, your CRM is broken.** AI follow-up tools are worthless if half your contacts have outdated emails or phone numbers from three moves ago. **Cleaning your database is not glamorous work, but it is the foundation on which everything else depends.**

Next, audit your response times honestly. When a lead comes in from your website or a listing portal, how long does it take you to respond? Not your best day, your average day. **Studies show that responding to real estate leads within five minutes dramatically increases conversion rates, but most agents take hours or even days.** AI can help with instant responses and lead qualifications, but you need to measure your current performance first. If you do not know how slow you are, you cannot calculate whether an AI response system is worth the monthly fee.

Then look at your market analysis process. When you prepare a comparative market analysis for a listing presentation, where does your data come from? How long does it take? Are you pulling comps manually from the MLS, or do you have systems that automatically aggregate neighborhood trends? AI valuation tools can accelerate this dramatically, but they need clean data inputs and human interpretation to be useful. **A Zestimate is not a pricing strategy—it is a starting point that requires your expertise to contextualize.**

For property managers, the preparation work is different. Your tenant communication records, maintenance request history, and lease renewal patterns contain gold that AI can mine for insights. **But only if that information exists in searchable form.** If your maintenance requests live in text messages and your lease data sits in filing cabinets, digitizing those records comes before any predictive maintenance AI.

Finally, think about what happens when technology fails. Real estate transactions have hard deadlines. If your AI scheduling system crashes the day before closing, do you have a backup plan? If your automated showing confirmation tool goes down during your busiest weekend, can your team handle it manually? **The agents who succeed with AI are the ones who treat technology as a powerful addition to their toolkit, not a replacement for knowing how to run their business without it.**

Your market knowledge and client relationships are still your competitive advantage. AI just helps you use them at scale.

Advance: Putting AI to Work

Real Estate Practice Implementation

The 90-day implementation demonstrates how real estate practices can achieve operational excellence while strengthening client relationships and market expertise:

30-Day Foundation Period:

- Lofty, formerly Chime CRM with AI features deployed across his 3-agent team.
- **AI-powered lead scoring reduced initial client qualification time** from 45 minutes down to 12 minutes.
- Automated nurture campaigns maintained contact with 127 prospective clients without manual follow-up.
- Client communication efficiency increased 52% through automated updates and scheduling.

60-Day Capability Expansion:

- ***Follow Up Boss*** integration for behavioral analysis and communication optimization.
- Advanced market intelligence through Top Producer AI-enhanced comparative analytics.
- Client portal **enhancement for real-time property alerts and market updates.**
- Transaction coordination improved through AI-assisted documentation and deadline tracking.

90-Day Strategic Integration:

- Custom workflows developed for different client types and transaction complexity levels.
- Advanced **analytics established for identifying referral opportunities** and client retention patterns.
- Comprehensive training programs created for ongoing AI capability development.
- Market positioning enhanced as a technology-supported real estate practice.

Complete Results:

- **Revenue growth of 134%** through expanded client capacity and premium service positioning.
- Client satisfaction increased 71% through more comprehensive market analysis and proactive communication.
- **Agent productivity increased 82%**—quality of personal consultations stayed high.
- 41% market share growth in the competitive East Valley market through technology-enhanced capabilities.

Three Best AI Programs for Real Estate Professionals

1. Lofty, formerly *Chime CRM with AI Lead Intelligence - The Complete Real Estate Business Brain*

What it does: AI-powered lead scoring, automated nurture campaigns, predictive client analytics, transaction management, and comprehensive business intelligence designed specifically for real estate professionals.

Why real estate agents choose it: ***Lofty*** provides intelligent lead management with automated follow-up systems that maintain client relationships while agents focus on active transactions.

Real impact: *A real estate* firm in Austin, Texas implemented ***Lofty's AI*** features and increased lead conversion rates by 89% while reducing transaction management time by 67%.

Monthly cost: $499-1500, depending on features and agent count.

Best for: Growing real estate teams that need comprehensive lead management with AI-enhanced client intelligence.

2. *Follow Up Boss with AI Behavioral Analysis* - The Client Relationship Optimizer

What it does: Smart lead distribution, behavioral analysis, automated communication sequencing, performance analytics, and team coordination, specifically designed for real estate sales teams.

Why relationship-focused agents adopt it: Follow Up Boss combines CRM functionality with AI insights that help agents understand client behavior and optimize communication timing.

Real impact: *Real Estate* Agency in Phoenix, Arizona, used Follow Up Boss's AI behavioral analysis to improve lead response times by 78% and increase listing appointments by 56%.

Monthly cost: $69-$1,000, depending on features and team size.

Best for: Real estate teams focused on systematic lead nurturing and client relationship development.

3. *Top Producer with Market Intelligence* - The Local Expertise Amplifier

What it does: AI-enhanced market analysis, comparative market analytics, client portal integration, transaction coordination, and neighborhood intelligence compilation.

Why market-focused agents choose it: Top Producer combines comprehensive market data with AI analysis to provide insights that enhance agent expertise and client consultation.

Real impact: Real estate professionals report 45% improvement in listing accuracy and 67% increase in client satisfaction through enhanced market intelligence and consultation quality.

Monthly cost: $80-200, depending on features and market access.

Best for: Individual agents and small teams focused on market expertise and client consultation excellence.

Connect: Making AI and Real Estate Expertise Work Together

True integration for real estate occurs when AI analytical power combines with human expertise to create service capabilities that neither could achieve alone.

The Complete Real Estate Integration Pattern

Every successful real estate AI implementation follows the same integration pattern:

AI Handles: Market data processing, lead scoring, automated follow-up, comparative market analysis, schedule coordination, and routine client updates.

Human Expertise Provides: Strategic pricing guidance, negotiation strategy, emotional support, complex problem-solving, relationship management, and ethical considerations.

Working Together Results: Provides comprehensive market analysis with personalized guidance, strategic insights from complete data, and quicker service without compromising trust.

Real-World Integration Success

Listing Pricing Enhancement:

AI Analysis: The system analyzed 284 comparable sales in the Chandler neighborhood for the Walker family home, identifying pricing trends, days-on-market patterns, and seasonal fluctuations to suggest an optimal listing price range.

Human Expertise: David interpreted AI recommendations within the context of the family's timeline pressure, the property's unique features, and the current buyer behavior patterns he observed during recent showings in the area.

Enhanced Result: Strategic pricing plan that achieved full-price offer within 8 days while addressing family-specific concerns about quick sale and maximizing value, leading to three referrals from impressed neighbors.

Buyer Client Management Integration:

AI Monitoring: Continuous analysis of new listings, price changes, and market conditions across 47 target neighborhoods with automated alerts matching the family's specific criteria and budget parameters.

Strategic Guidance: David utilized AI insights to provide proactive recommendations based on the family's commute needs, school priorities, and long-term investment goals that algorithms cannot assess.

Superior Outcome: 67% reduction in unnecessary property showings while identifying an ideal home within 3 weeks, increasing client satisfaction through an efficient process, and strengthening the reputation for understanding client needs beyond basic search criteria.

Keep: Maintaining Real Estate Excellence

Real estate **professionals who successfully implement AI create advantages that strengthen over time.** They do this by establishing systems that enhance market expertise while preserving the trust relationships that define their value.

Industry Recognition and Growth

Larry's Real Estate Recognition:

- The company was **featured in real estate publications** as a model for integrating AI efficiency with comprehensive client service.

- Larry was presented with opportunities to present at brokerage conferences on the topic of responsible AI implementation within the real estate sector.

- The company enjoyed an **increase in client referrals by 83%** due to improved service capabilities and transaction outcomes.

Professional Development Through Technology

- Position **AI implementation as a commitment to client service excellence,** not merely cost reduction.

- Leverage technological capabilities to deliver more advanced market analysis and strategic insights.

- Show that AI cannot substitute for professional judgment or relationship management.

- **Build a reputation as an agent who combines cutting-edge capabilities with trusted guidance.**

The Real Estate Advantage

The most successful real estate professionals do not treat AI as a replacement for market knowledge or a shortcut around relationship building. They use it as a business amplifier that requires experienced agents to fully realize its potential.

Key Success Elements

- **AI handles lead qualification,** market data processing, and routine follow-up communication.
- **Agents provide strategic pricing guidance,** negotiation expertise, and emotional support during major life decisions.
- Client **relationships remain personal** while service capabilities expand to match client expectations.
- Market knowledge deepens as AI surfaces patterns that manual analysis would miss.

Real estate excellence is not about choosing between personal service and digital efficiency. **It is about using AI to make experienced agents more responsive, more informed, and more valuable than ever before.**

Real estate professionals thoughtfully using AI create advantages that discount brokerages and algorithm-only platforms cannot replicate. **The combination of instant responsiveness, comprehensive market intelligence, and trusted personal guidance takes years to develop.** A Zillow estimate is just a number. An agent who can explain why that number is wrong for this specific house, on this specific street, given what the neighbors sold for and what buyers in this price range actually care about? That is expertise worth paying for.

The *DeBerry Team* did not become less personal when they added an AI assistant. They became a team that never misses a lead at midnight and does not lose relationships because someone was too busy showing houses to return a call. Roxanne still sits at kitchen tables. Larry still runs the marketing. The AI just makes sure opportunities do not slip away while they are doing the work that actually closes deals.

Here is what should keep real estate professionals who wait to implement AI up at night: While you are deciding whether a CRM upgrade is worth the monthly fee, your competitor down the street just responded to that online lead in forty-five seconds. You responded in four hours. By the time you called, a showing had already been scheduled.

The agent who implemented AI lead scoring six months ago is now spending her mornings with qualified buyers instead of chasing dead ends. The property manager who digitized his maintenance records caught three potential problems before tenants complained. **They are not technology companies. They are real estate professionals who decided that "I will get back to you" was no longer good enough, given that the competition gets back to them instantly.**

Your clients have options. Every listing presentation, every buyer consultation, every lease renewal is a moment where someone else could earn their business. The relationships you have built over years matter, but **relationships do not survive indefinitely when service quality falls behind expectations.** The client who loves you will still take a meeting with the agent who sent them a personalized market update before you remember to call.

The technology is not complicated. The tools exist. The only question is whether you will use them to become the agent who combines decades of market knowledge with the responsiveness that modern clients demand, or whether you will keep telling yourself that relationships alone will protect you from competitors who offer relationships plus everything else.

Your market expertise is not becoming obsolete. Your negotiation skills are not being automated. Your ability to guide families through the most significant financial decisions of their lives is not something an algorithm can replicate. But the agents who wrap that expertise in AI-enhanced service will win the listings. They will convert the leads. They will build referral networks that compound over time.

Roxanne DeBerry started selling homes in 1975. Fifty years later, her team uses artificial intelligence that would have seemed impossible when she made her first sale. **The fundamentals have not changed. Relationships still drive everything.** AI just makes sure no opportunity slips through the cracks while you are busy serving the clients already in your care.

How long before your best clients start wondering why the other agent always seems one step ahead?

Chapter 23

Financial & Accounting Services

Your clients trust you with their financial future, their tax obligations, their insurance protection, and their biggest purchases. That trust is not negotiable.

Financial and accounting professionals face a unique challenge: how do you **use AI to serve more clients better,** without losing the deeply personal relationships that make clients choose you over automated alternatives? The answer is not replacing human judgment with algorithms. The answer is **using AI to handle time-consuming technical work** so you have more time for the strategic conversations that clients actually value.

This chapter covers **financial and wealth advisory, accounting services, insurance agents, mortgage brokers, and tax advisors**, each with unique challenges and specific AI solutions. Whether you advise on investments, prepare taxes, sell insurance, facilitate mortgages, or manage accounting, you will find practical implementation guidance for your particular practice.

Success Story
Carson Group

Carson Group built its reputation on high-touch financial planning and deep client relationships. But as the firm grew, advisors found themselves buried under a **mountain of administrative work** that had nothing to do with actually helping clients.

Every client meeting triggered the same exhausting cascade: drafting summary emails, updating CRM records, documenting financial goals, logging concerns, capturing next actions, noting personal details, and preparing handoffs. None of this work generated revenue. **None of it strengthened relationships**. But skip the documentation, and you are looking at compliance gaps, inconsistent service, and operational nightmares.

The numbers told the story. A**dvisors were losing eight to twelve hours per week to manual note-taking** and follow-up. Across the entire firm, that added up to more than 18,000 hours annually disappearing into paperwork.

Leadership knew something had to change, but traditional AI meeting tools created their own problem: clients had to be recorded. Many clients felt uncomfortable knowing their financial conversations were being captured digitally, and advisors did not love the disclosure requirements either.

Then they found ***Zocks***, which does something genuinely different. It listens only to the advisor's side of the conversation and converts spoken notes into structured meeting summaries, follow-up action items, CRM entries, and relationship details. The advisor does not type a word. The client never feels monitored. Compliance stays clean.

After a small pilot impressed even the skeptics, *Carson Group* rolled it out firmwide. **The results were immediate and measurable.** Within six months, the firm recaptured 9,000 hours of advisor time, the equivalent of four and a half full-time advisors without adding payroll. The measurable return on investment exceeded two million dollars. Advisor adoption hit 92 percent, and every meeting now had the same level of documentation regardless of advisor style.

But here is what matters most: freed from their laptops, advisors gave clients their full attention. They listened more deeply, **asked better questions, and picked up nuances that would have been missed while typing.**

One *Carson* advisor described it perfectly: "It did not make me more efficient. It made me more present. Clients feel it. I feel it."

Within 60 days, every client profile had comprehensive relationship details, follow-up tasks were executed with near-perfect accuracy, and advisors were spending evenings with their families instead of their laptops. **AI did not replace these advisors.** It amplified them and returned the firm to what built its reputation in the first place: human connection at scale.

Trust: Building Confidence in AI-Driven Financial Guidance

Financial professionals face two audiences who need trust-building: clients who worry AI will replace personal service, and team members who fear technology will eliminate their jobs.

Building Client Confidence

Clients need to understand that **AI handles routine tasks** while you focus on strategic guidance. Here is the conversation that works:

"We have implemented AI tools that handle time-consuming data entry, document processing, and routine calculations. This gives me more time for what you actually hired me for - analyzing your specific situation and providing personalized recommendations. You will get faster responses to routine questions and more thorough analysis of complex decisions."

Notice what this conversation does not say: it does not promise AI will make everything automated, it does not claim technology replaces expertise, and it does not suggest clients will interact with chatbots instead of professionals. You are clear that **AI enhances your service and does not replaces it.**

Building Team Confidence

Your team members need different messaging. They fear unemployment, not service quality. The conversation that addresses their actual concerns:

"These AI tools eliminate the repetitive work that keeps you from using your expertise. Data entry, document organization, routine calculations - AI handles those tasks so you can focus on client relationships and complex problem-solving. We are not replacing people. **We are eliminating the work that makes talented professionals spend their time on tasks computers should handle."**

Then prove it. Start with the tasks everyone hates: manual data entry from PDFs, document organization, appointment scheduling, follow-up reminders. When your team sees AI eliminating their least favorite work without threatening their jobs, resistance disappears.

Ready: Preparing Your Practice for AI Implementation

What keeps financial professionals up at night? The fear that one software glitch, one data breach, one compliance failure will destroy the trust they spent decades building. **That fear is not irrational.** When you handle other people's money, mistakes have consequences that go far beyond a bad online review. So, when someone suggests adding AI to your practice, your instinct to proceed carefully is exactly right.

Here is what the eager technology vendors will not mention: **financial services AI fails spectacularly when it meets sloppy data.** That client management system you have been meaning to clean up for three years? The one with duplicate records, outdated addresses, and notes that only make sense to the person who wrote them? AI will amplify every bit of that mess.

Before you implement any tool that promises to automate client communications or generate financial reports, you need to know exactly what shape your data is in. Pull a random sample of fifty client records and check them against reality. How many of them have current contact information? Complete financial profiles? Accurate service histories? **If that number is below 90% then data cleanup is your first project, not AI implementation.**

Next is the compliance question, and it cannot be stressed enough that **your regulators do not care if AI made the mistake.** If an automated system sends incorrect tax information to a client, it generates a misleading financial projection, or misfiles a compliance document, that is your license on the line. **Every AI tool in your practice needs a human checkpoint before anything goes to a client or a regulatory body.** We have watched firms get excited about "fully automated" workflows only to realize they had removed the professional review step that kept them out of trouble. Build your quality control process before you build your automation.

Talk with your errors and omissions insurance carrier before you implement anything. Some carriers have specific requirements about AI-assisted professional services, and discovering a coverage gap after something goes wrong is not a conversation you want to have. Ask specifically about AI-generated client communications, automated financial calculations, and document preparation tools. **Get the coverage in writing.**

Your team preparation matters more in financial services than almost any other industry. **Position AI as a first draft that experienced professionals review, not a replacement for their judgment.** When your team understands they are supervising the AI rather than being replaced by it, resistance transforms into constructive feedback about what the system gets wrong.

Finally, know your clients. Some will love hearing that you have invested in technology to serve them better. Others may worry that "AI" means less personal attention and more automation. **Segment your client communication accordingly.** Your tech-forward clients get the message about efficiency. Your relationship-focused clients get reassurance that technology means you have more time for the strategic conversations they value. Same implementation, different framing, based on what each client actually cares about.

The financial practices that succeed with AI are the ones that remember what they are actually selling: trust, judgment, and expertise. Technology should make those things more powerful, not replace them.

Advance: 90-Day Implementation

Financial services span diverse specialties, each requiring different AI tools and implementation approaches. This section provides specific guidance for five major categories.

Financial Advisors Implementation Timeline

Month 1: Foundation & Testing

- Implement meeting intelligence tools on internal team meetings first.
- **Test automated portfolio reporting** with 10 pilot clients.
- Document time savings on routine client communications.
- **Build template library** for common client scenarios.

Month 2: Client Rollout

- **Apply AI tools to 50% of client base** for meeting preparation.
- Implement automated portfolio performance reporting for all clients.
- Use AI-generated summaries for complex financial planning scenarios.
- **Track time savings and** client satisfaction metrics.

Month 3: Full Integration

- **All client interactions supported by AI-enhanced preparation.**
- Automated alerts for portfolio rebalancing opportunities.
- AI-driven identification of tax-loss harvesting opportunities.
- Measure capacity increase and client engagement improvements.

Three Best AI Tools for Financial Advisors

1. Zocks AI Meeting Assistant and CRM Updates for Advisors

What it does: Captures client meeting details without recording, automatically generates follow-up action items, maintains comprehensive client profiles, and integrates with CRM systems. Designed specifically for financial advisors who need meeting intelligence without compliance concerns.

Why advisors choose it: The technology captures relationship details, financial goals, and action items without requiring clients to know they are being monitored. You focus on client conversations while AI handles documentation.

Real impact: *Carson Group* saved 9,000 advisor hours in six months with over $2 million ROI. *RFG Advisory* eliminated manual meeting notes entirely while improving client relationship documentation.

Monthly cost: $67-220/advisor, depending on features.

Best for: Financial advisors and wealth managers who conduct regular client meetings and need better relationship intelligence without recording conversations.

60-day milestone: Your client profiles contain comprehensive relationship details accumulated automatically from meetings, with AI-generated action items that eliminate follow-up delays.

2. *eMoney Advisor* - Comprehensive Financial Planning and Scenario Modeling

What it does: Comprehensive financial planning platform with AI-enhanced analysis, client portal for document sharing, scenario modeling for retirement and education planning, and automated what-if analysis.

Why advisors choose it: The platform handles complex financial planning calculations while presenting results in client-friendly visualizations. You demonstrate value through comprehensive analysis without spending hours building spreadsheets.

Real impact: Financial advisors using ***eMoney*** report 40% reduction in plan preparation time while delivering more sophisticated analysis. Client engagement with planning portal averages 8-12 interactions monthly.

Monthly cost: $200-400/advisor depending on client count and features

Best for: Comprehensive financial planners who need sophisticated analysis tools and client-facing planning portals.

60-day milestone: Your financial planning presentations include comprehensive scenario analysis that would have required days of manual calculation, delivered in formats clients actually understand.

*3. Nitrogen (previously **Riskalyze**): **Risk Tolerance Platform Measuring Investor Risk***

What it does: Risk tolerance assessment using behavioral analysis, portfolio alignment scoring, automated rebalancing recommendations, and client-friendly risk reporting.

Why advisors choose it: The technology quantifies client risk tolerance through proven behavioral assessment, eliminating subjective guesswork. You align portfolios with actual client comfort levels rather than assumed preferences.

Real impact: Advisors report 60% reduction in panicked client calls during market volatility because portfolios match documented risk tolerance. Client retention during market downturns improves significantly.

Monthly cost: $250-399/advisor, depending on assets under management.

Best for: Investment advisors managing portfolios who need objective risk assessment and alignment verification.

60-day milestone: Every client portfolio has a documented risk tolerance score with automated monitoring for drift, reducing your compliance exposure while improving client outcomes.

Three Best AI Tools for Insurance Agencies

1.Ezlynx With Eva Virtual Assistant Ezlynx: AI-Enhanced Agency Management With Automated Workflows

What it does: Comparative rating across multiple carriers, automated client communication via text and email, account summarization from communication history, and AI-powered email content assistance.

Why agents choose it: The platform handles multi-carrier quoting and routine client service automatically, making small agencies competitive with captive carriers on convenience.

Real impact: Independent agencies report 60% reduction in time spent on routine client service requests. EVA responds to client texts about ID cards and coverage details automatically.

Monthly cost: $150-300/user, depending on features and agency size

Best for: Independent insurance agents who need multi-carrier comparative rating and automated client service.

60-day milestone: Your agency provides automated responses to routine client requests 24/7, with AI summarizing client communication history before you return calls.

2. *Applied Epic* – Property & Casualty Management

What it does: Commercial insurance agency management with AI-powered account rounding recommendations, automated statements reconciliation, risk assessment analysis, and workflow automation.

Why agents choose it: The platform identifies cross-selling opportunities automatically and reduces commission reconciliation from 15 hours weekly to minutes.

Real impact: Commercial agencies using ***Applied Epic*** report significant efficiency gains in back-office operations with automated statement processing and reconciliation.

Monthly cost: $200-500/user for commercial insurance features

Best for: Commercial insurance agencies tired of manual commission reconciliation and seeking intelligent account rounding recommendations.

60-day milestone: Your accounting staff spends minutes instead of hours on commission reconciliation, and the system flags cross-selling opportunities automatically.

3. HawkSoft --Agency Management System

What it does: Agency management system with integrated comparative rating, automated workflows, client portal, and document management.

Why agents choose it: The platform combines agency management with rating capabilities in one system, eliminating the need for separate management and rating platforms.

Real impact: Agencies report consolidated workflows with faster quote-to-bind cycles and improved client service through self-service portals.

Monthly cost: $125-250/user, depending on features

Best for: Independent agencies wanting an all-in-one management and rating solution.

60-day milestone: Your agency operates from a single platform for management and quoting, with clients accessing their documents through a self-service portal.

Mortgage Brokers & Loan Officers

Success Story
Riverbank Finance

Anthony Bird and Joshua Woodwyk started *Riverbank Finance* in Grand Rapids, Michigan in 2011, right after the mortgage industry had been turned upside down.

Every mortgage broker knows the drill. You meet with buyers, qualify them, find them the right loan product, then spend the next several weeks sending emails and making phone calls asking for pay stubs, bank statements, tax returns, and the dozen other documents underwriters demand. Then you ask again when something is missing a page. Then you ask a third time when the borrower sends an outdated version.

Riverbank had a full-time processor whose entire job was document collection. When she left the company, Bird and his partner **faced a decision:** hire another person to do the same frustrating work, or find a better system. They chose ***Floify***.

"When you have a person collect documents, there are organizational issues and communication issues where clients aren't really sure who this third-party person is," Bird explains. "Borrowers might not get instant feedback if they send a bank statement and it's missing a page. Our staff might not get to it for a few days." **The old system created friction for everyone involved.**

Floify gave their borrowers a secure portal where they could upload documents directly, see exactly what was still needed, and get automated reminders. The system checked documents instantly and told borrowers when something was incomplete. No more phone tag. No more buried emails. No more processors spending their days being the squeaky wheel.

The transformation was immediate and measurable. *Riverbank Finance* now closes loans in half the time of the industry average. Not slightly faster. Half.

"Speed is a huge priority for us," Bird says. "We know that if we can close loans faster, we're going to impress more clients and realtors and earn more referral business."

Here is what makes this story relevant for other small brokers: *Riverbank* did not add staff to grow. **They grew because their existing team could handle exponentially more loans.** Bird describes the automation as "the equivalent of having an assistant for each one of our employees."

The client experience improved as much as the internal operations. Before, borrowers had to wait for someone to get back to them about what was needed. Now they log into a portal and see exactly where their loan stands. The system sends milestone updates automatically, so clients and their realtors know what is happening before they even think to ask.

"***Floify*** is a huge tool for communication as much as it is for gathering documents," Bird notes. "We use it from start to finish."

Riverbank Finance did not adopt technology to look impressive or because everyone else was doing it. They adopted it because chasing documents was stealing time from actual mortgage work, and they found a way to get that time back. The result is a small brokerage that competes with big lenders on speed and service, without the big lender overhead.

Three Best AI Tools for Mortgage Brokers

1. *Floify -- Mortgage POS* Automating Documents, Disclosures, and Communications

What it does: Automated document collection from borrowers, secure client portal for application and document upload, automated communication and reminders, and integration with loan origination systems.

Why brokers choose it: The platform eliminates document chasing through email and provides borrowers with a professional, branded experience that matches big lender platforms.

Real impact: Mortgage brokers report 60-70% reduction in time spent collecting and organizing borrower documents. Loan processing cycles drop by 30-40%.

Monthly cost: $150-400, depending on loan volume

Best for: Independent mortgage brokers tired of chasing borrowers for documents via email.

60-day milestone: Your borrowers upload documents to a secure portal with automated reminders, eliminating your manual document collection process.

2. *Blend* -- Digital Lending Platform Unifying Mortgage Application to Close

What it does: Digital mortgage application platform with automated document verification, income and asset verification, real-time application status, and e-signature capabilities.

Why brokers choose it: The technology provides consumer-grade digital experience that matches big bank platforms—with broker flexibility intact.

Real impact: Brokers using ***Blend*** deliver digital-first experience competitive with national lenders, reducing application abandonment rates by 40-50%.

Monthly cost: Custom pricing based on loan volume

Best for: Mortgage brokers wanting enterprise-grade digital application experience.

60-day milestone: Your borrowers complete applications on mobile devices with automated document verification, matching the convenience of national lender platforms.

3. *Mortgage Coach* -- Visual Loan Presentations Transforming Rate Conversations Into Advice

What it does: Interactive mortgage presentation platform with scenario comparison, total cost analysis visualization, and client-facing calculators for informed decision-making.

Why brokers choose it: The technology helps borrowers understand complex loan options through visual comparisons, positioning you as advisor rather than order-taker.

Real impact: Brokers report higher borrower confidence in loan decisions and reduced last-minute program changes. Client education becomes consultative conversation rather than spreadsheet review.

Monthly cost: $99-299/month depending on features

Best for: Mortgage brokers who want to position themselves as financial advisors helping borrowers make informed decisions.

60-day milestone: Your loan presentations include visual scenario comparisons that help borrowers understand total cost implications, positioning you as trusted advisor.

Connect: Human Expertise + AI Efficiency = Client Value

Financial services succeed when AI handles technical work while humans provide judgment, strategy, and relationship management. This section shows how each service category integrates AI with human expertise.

Financial Advisors: Strategic Partnership Model

AI Handles: Meeting documentation, client profile maintenance, portfolio performance calculations, rebalancing recommendations, risk tolerance assessment.

People Provide: Strategic financial advice, life planning conversations, behavioral coaching during market volatility, complex estate planning, and relationship management.

Working Together Results: Advisors spend 70% of client meetings on strategic planning rather than performance review. Client satisfaction increases because conversations focus on goals rather than data.

Insurance Agencies: 24/7 Service Capability

AI Handles: Routine client inquiries, ID card requests, multi-carrier rating, automated communication, policy documentation.

People Provide: Coverage analysis, claims advocacy, complex risk assessment, relationship management, policy explanation.

Working Together Results: Agencies provide instant responses to routine requests—with personal service for important decisions. Client retention improves through better availability.

Mortgage Brokers: Competitive Digital Experience

AI Handles: Document collection, application processing, status updates, automated reminders, compliance documentation.

People Provide: Program selection guidance, complex scenario navigation, lender relationship leverage, closing coordination, personalized advice.

Working Together Results: Brokers deliver big-bank digital convenience with personalized service advantage. Processing time decreases while borrower satisfaction increases.

Keep: Maintaining Competitive Advantage

AI implementation delivers immediate efficiency gains, but sustained competitive advantage requires continuous improvement and strategic positioning.

Six-Month Success Metrics

- Client **capacity increased 30-40%** without adding staff.
- **Response time** to client inquiries **decreased by 60-70%.**
- Time spent on routine tasks reduced by 50-65%.
- Client satisfaction scores improved across all service categories.
- **Team members report reduced stress** during busy seasons.
- Revenue per team member increased 25-35%.

Continuous Improvement Strategies

- Conduct a Monthly AI Performance Review of your AI applications.
- Analyze which automated tasks require frequent human correction.
- **Identify new workflow bottlenecks** that AI could address.
- Review client feedback on automated versus personal service.
- **Assess team adoption and identify training needs.**
- Compare productivity metrics against pre-AI baseline.

Strategic Market Positioning

Your competitors divide into two categories: **those using AI to cut prices through automation,** and those maintaining traditional service models at traditional costs. You occupy the valuable middle ground: AI-enhanced service at competitive prices with superior client experience.

Market this advantage explicitly. Y**our clients should understand they receive both modern convenience and personal expertise,** not choosing between them.

Team Development for AI-Enhanced Practice

- **Train team members on strategic advisory skills** as routine work decreases.
- **Reward client relationship quality rather than transaction volume.**
- Develop specialized expertise areas that AI cannot replicate.
- **Create career progression paths** focused on client advisory rather than task completion.
- Measure success by client outcomes rather than hours worked.

The Financial Services Advantage

The most successful financial professionals **do not treat AI as a shortcut around compliance** or a replacement for fiduciary judgment. They use AI as a practice amplifier that requires experienced advisors to fully realize its potential.

Key Success Elements:

- **AI handles meeting documentation, data entry, document collection,** and routine client communications.
- **Professionals provide strategic advice,** behavioral coaching, complex problem-solving, and relationship management.
- Client relationships remain personal while service capabilities expand to meet modern expectations.
- **Compliance stays tight** because humans review everything before it reaches clients or regulators.

Financial services excellence is not about choosing between personal attention and digital efficiency. **It is about using AI to make experienced professionals more present, more thorough, and more valuable than ever before.**

Financial practices intentionally using AI create advantages that robo-advisors and automated platforms cannot replicate. **The combination of instant responsiveness, comprehensive analysis, and trusted personal guidance takes years to develop.** A portfolio algorithm is just math. An advisor who can talk a panicked client off the ledge during a market downturn, who remembers that the Johnson family needs to fund college in three years, who understands why this particular widow is terrified of risk even when the numbers say she can afford it? That is expertise no chatbot can fake.

Carson Group did not become less personal when they implemented AI meeting assistants. They became a firm where advisors actually look clients in the eye instead of typing notes. They recaptured 9,000 hours in six months, not by working less, but by working on what matters. *Riverbank Finance* did not lose their personal touch when they automated document collection. They started closing loans in half the industry average time, and their borrowers stopped dreading the paperwork phase entirely.

Here is what should keep financial professionals who wait to implement AI up at night: While you are deciding whether a meeting assistant is worth the monthly fee, your competitor across town just documented every client conversation automatically and followed up the same day. You are still typing notes at 9 PM. They are having dinner with their family.

The insurance agent who implemented AI six months ago responds to routine client requests in seconds, not hours. The mortgage broker who automated document collection closes loans while you are still chasing pay stubs through email. The wealth advisor who deployed risk tolerance software has documentation that will hold up under regulatory scrutiny. Yours is a conversation you half-remember from last quarter. **They are not technology firms. They are financial professionals who decided that "I will get back to you"**

was no longer good enough when clients expect instant answers from everyone else.

Your clients have options. The robo-advisors are not going away. The online lenders keep improving. The direct-to-consumer insurance platforms get slicker every year. **The relationships you have built over decades matter, but relationships do not survive indefinitely when service quality falls behind what clients experience everywhere else.** The client who trusts you with their retirement will still take a meeting with the advisor who sent a comprehensive portfolio review before you remembered to call.

The compliance requirements are not getting simpler. The client expectations are not getting lower. The competitors are not getting slower. **The only question is whether you will use AI to become the professional who combines decades of expertise with the responsiveness that modern clients demand, or whether you will keep telling yourself that trust alone will protect you from firms that offer trust plus everything else.**

Your judgment is not being automated. Your fiduciary responsibility is not being outsourced. Your ability to guide families through the most consequential financial decisions of their lives is not something an algorithm can replicate. But the professionals who wrap that expertise in AI-enhanced service will win the clients. They will earn the referrals. They will build the practices that compound over time while everyone else wonders where the business went.

One *Carson* advisor said it perfectly: "It did not make me more efficient. It made me more present. Clients feel it. I feel it." That is what AI implementation looks like when you do it right. Not replacing what makes you valuable. Amplifying it.

How long before your best clients start wondering why the other advisor always seems to have more time for them?

Chapter 24

Professional Services

Showcasing Intelligence That Clients Value

Your professional expertise is not outdated; it is growing stronger.

Professional services succeed through the combination of human expertise and client trust cultivated over time. **Lawyers**, **consultants**, **architects**, and other **knowledge professionals** face a unique challenge: clients hire you for your judgment, experience, and strategic thinking. Capabilities that seem fundamentally human.

Yet **AI-enhanced competitors are delivering faster research, more comprehensive analysis, all at lower costs** that threaten traditional practice models. The AI transformation is not about becoming a technology company; it is about becoming a more powerful professional who uses AI to deliver insights and value that human-only approaches cannot match.

Success Story
Immigration Law

*David** the immigration lawyer, built his immigration law practice on a foundation of personal attention and deep expertise in navigating complex visa cases. His clients valued the way he took time to understand their unique circumstances, explained legal options clearly, and guided them through stressful processes with patience and care.

But major law firms were targeting his client base with **AI-enhanced services that promised expedited case processing and lower fees.** These larger competitors could generate documents in hours rather than days, provide 24/7 client communication through automated systems, and undercut his pricing while handling higher case volumes.

The professional services dilemma: How do you use AI to deliver faster, more comprehensive analysis without losing the human judgment and personalized guidance that clients hire you for in the first place?

Trust: Building Professional and Client Confidence

Professional service clients hire you for judgment that they believe AI cannot provide. The discussion about trust involves framing AI as a tool that enhances the scope and responsiveness of your expertise instead of replacing it.

Scripts for Building Client Trust

For clients concerned about AI replacing human judgment: "AI helps me research more thoroughly and analyze more data than ever before, but you are still getting my experience, my strategic thinking, and my judgment on your specific situation."

For clients worried about confidentiality: "All AI tools we use are specifically designed for professional services with the highest security standards. Your information remains completely confidential, and AI just helps me serve you more efficiently."

For clients questioning value: "AI handles the routine research and documentation that used to take hours, so I can spend more time on strategic analysis and providing the advisory services you value most."

Professional Services Trust Plans

Every successful professional service AI implementation requires three trust conversations:

With Senior Staff: Position AI as a career advancement opportunity that makes senior expertise more valuable through enhanced analytical capabilities.

With Junior Staff: Frame AI proficiency as an essential modern professional skill that accelerates career development and increases value to clients.

With Clients: Emphasize AI as a tool for more comprehensive analysis and faster service delivery while maintaining personal attention and expert judgment.

Ready: Preparing Systems and People

Professional services live and die on reputation. A lawyer who misses a filing deadline, an architect whose design fails inspection, a consultant who gives advice that tanks a client's project—these are not recoverable errors. Your clients hire you because they believe your expertise protects them from expensive mistakes. So, when AI vendors promise to revolutionize your practice, your skepticism is not resistance to change; it is professional responsibility.

Here is the preparation that separates **successful AI implementation from expensive embarrassment**.

Start by mapping where your time actually goes. Not where you think it goes, where it actually goes. Track your hours for two weeks with brutal honesty. How much time do you spend on work that requires your professional judgment versus work that a well-trained assistant could handle if you had one? **Most professionals are shocked to discover that 40-60% of their billable hours go to tasks that do not require their expertise:** document formatting, research compilation, scheduling coordination, status updates. That is your AI opportunity. But you cannot optimize what you have not measured.

Next, audit your document situation. Professional services run on documents: contracts, briefs, reports, proposals, specifications. Where do yours live? If the answer involves multiple computers, various cloud services, email attachments, and "somewhere on the shared drive," you have a problem AI cannot solve. **AI document tools need organized, searchable archives to be useful.** If finding last year's version of a standard contract requires archaeology, fix your filing system before you add technology on top of it.

Then get serious about confidentiality. Your clients trust you with sensitive information on legal matters, financial details, strategic plans, and personal circumstances. Before any AI tool touches that data, you need to understand exactly where the information goes, who can access it, and what happens to it after processing. Read the privacy policies. Ask vendors direct questions about data handling. **Some AI tools train on user inputs,** which means your client's confidential information could theoretically influence outputs for other users. That may be acceptable for some applications and absolutely unacceptable for others. Know the difference before you implement.

Professional licensing boards are watching AI adoption closely, and the rules are still evolving. What constitutes appropriate AI assistance versus unauthorized practice? When does AI-generated work product require disclosure to clients? How do professional responsibility rules apply to automated advice? These questions do not have universal answers yet, but your state bar, licensing board, or professional association likely has guidance—or is developing it. **Check before you implement and document your compliance approach.**

Finally, have an honest conversation with yourself about what makes you valuable. If your competitive advantage is doing routine work faster and cheaper than alternatives, AI is a threat because it can do that work faster and cheaper than you can. If your advantage is judgment, relationships, and expertise that clients cannot get from a software platform, **AI is an opportunity to do more of what makes you irreplaceable.** The professionals who thrive will be the ones who use AI to eliminate the commodity work so they can focus on the premium work that justifies their fees.

Your expertise is not obsolete. But **expertise alone is no longer enough.** The question is whether you will use AI to make your judgment more powerful, or watch competitors do it first.

Advance: Putting AI to Work Across Professional Services

Professional services firms are discovering that AI is not about replacing the expertise they have built over decades, but supplementing it in ways that free their teams to focus on what actually matters - serving clients. **The AI tools are already here, built into the software these businesses use every day,** quietly handling the document sorting, data entry, and routine communications that used to eat up billable hours.

Legal Services Excellence

The immigration law firm implemented AI within its practice and streamlines document automation, legal research, and reviewing case histories to manage its practice operations and client communication.

Results after 12 months:

- Case **completion time decreased by 52%** due to automated document preparation.
- Client **satisfaction rose 45%** due to better communication and more accurate timelines.
- **Revenue grew by 89%** due to higher capacity and premium positioning.
- Market differentiation identified for an immigration law firm that uses technology and offers a personalized service.

Three Best AI Programs for Legal Services

1. *Clio Legal AI* - The Complete Legal Practice Brain

What it does: AI-powered practice management, client communication automation, billing optimization, calendar integration, and legal workflow improvement.

Why legal professionals choose it: Clio is explicitly designed for law firms with industry-specific workflows and compliance requirements.

Real impact: Immigration law practices report a 52% reduction in case completion time while improving client communication and satisfaction.

Monthly cost: $49-149 per user per month for Clio Manage, with most law firms paying $89-149 once they add features for intake, automation, and AI capabilities. Manage AI (formerly Clio Duo) is included in current plans.

Best for: Solo practitioners and small law firms who are focused on client service and case management efficiency.

2. *CaseText with AI Research* - The Legal Intelligence Engine

What it does: ***CaseText AI*** is powered by legal research across comprehensive legal databases and includes case analysis, brief writing assistance, and legal precedent identification.

Why attorneys choose it: The program offers advanced legal research tools powered by AI insights that are designed to complement, not substitute, professional legal expertise.

Real impact: Law firms report a 60% reduction in research time while improving the quality of their legal brief and thorough case preparation.

Monthly cost: $65/month per user.

Best for: Attorneys who are focused on complex legal research and case preparation across multiple practice areas.

3. *Gavel AI* - The Document Automation Workhorse

> **What it does:** AI-powered document automation platform that converts client intake questionnaires into finished legal documents, combining conditional logic, client portals, and AI contract review directly in Microsoft Word.
>
> **Why legal professionals adopt it:** The system eliminates repetitive drafting work that burns billable hours, letting solo practitioners and small firms handle volume that previously required multiple associates.

Real impact: A survey of 50 law firms showed a 90% reduction in document drafting time, turning five-hour divorce filings into 30-minute automated workflows and six-hour estate planning packets into an hour of work. Immigration firm *Sisu Legal* reports saving 75% of time per case while reducing errors.

Monthly cost: Plans range from $83 to $290 a month. AI contract review tool (*Gavel Exec*) runs $160/month per user for unlimited usage.

Best for: Estate planning, family law, real estate, and immigration practices drowning in repetitive document preparation, particularly firms wanting to create client-facing intake portals.

Accounting & Bookkeeping Services

Accounting firms are using AI to tackle the grunt work that has always stolen time from actual client advisory. With platforms like ***QuickBooks with Intuit Assist*** now auto-categorizing expenses, draft invoice reminders, and surface cash flow insights without anyone touching a spreadsheet. **Even document collection has gotten smarter, extracting data from photographed receipts and emailed invoices,** while AI automatically renames and tags every file clients upload, turning "IMG_4523.pdf" into a named file that a person can actually find.

Three Best AI Tools for Accounting & Bookkeeping

1. QuickBooks Online with AI Features Automating Invoices, Categorization, and Cash Flow

What it does: AI-powered transaction categorization, automated bank reconciliation, receipt capture with optical character recognition, cash flow forecasting, and expense tracking.

Why accountants choose it: The platform handles routine bookkeeping automation—with the flexibility accountants need for complex scenarios. You supervise AI categorization rather than doing it manually.

Real impact: Accounting firms report 50-60% reduction in transaction processing time. Month-end close cycles drop from 7-10 days to 2-3 days.

Monthly cost: $35-250 per client, depending on features and AI capabilities.

Best for: Bookkeepers and accountants serving small business clients who need automated transaction processing.

60-day milestone: Your team reviews AI-categorized transactions rather than categorizing them manually, with accuracy rates exceeding 85% requiring minimal corrections.

2. *Bill.com* AI-Powered AP/AR Automation Built for Accounting Firms and Small Businesses

What it does: Automates the entire accounts payable and receivable workflow—invoice capture, approval routing, payment scheduling, vendor payments, and customer invoicing. AI reads invoices, extracts data, suggests GL coding, and learns your approval patterns.

Why accountants choose it: Bill.com eliminates paper checks, manual data entry, and approval bottlenecks. The platform learns your vendors, coding patterns, and approval workflows—routing invoices to the right people automatically while flagging exceptions for review.

Real impact: Accounting firms using Bill.com process invoices 50% faster and reduce payment errors by 70%. Teams eliminate paper check runs entirely, with vendors receiving ACH or virtual card payments automatically on schedule.

Monthly cost: $45-79/month for small businesses; $79-149/month for accounting firms managing multiple clients (pricing scales with transaction volume and users).

Best for: Accounting firms managing AP/AR for multiple clients, or businesses paying 50+ bills monthly and tired of paper check chaos.

60-day milestone: Your clients' bills are approved and paid electronically without manual check writing. Invoice data flows automatically into QuickBooks with minimal data entry.

3. *HubDoc (Xero)* -- Snap, Scan, Extract—Bills and Receipts Into Xero Automatically

What it does: Automated document collection from suppliers, receipt and invoice scanning with data extraction, automated data entry into accounting systems, and organized document storage.

Why accountants choose it: The technology eliminates manual receipt entry and document chasing, with clients forwarding receipts automatically without your involvement.

Real impact: Firms report 70% reduction in time spent collecting and entering receipts and invoices. Month-end document collection drops from hours to minutes.

Monthly cost: $20/client with Xero subscription

Best for: Accountants tired of chasing clients for receipts and invoices every month.

60-day milestone: Your clients forward receipts to an email address where AI extracts data automatically, eliminating your manual document collection process entirely.

Tax Preparation & Planning Services

Tax preparers are watching **AI transform the most tedious parts of their workflow.** Software now recognizes uploaded W-2s and 1099s instantly, renames files automatically, and even suggests tax-saving strategies by analyzing patterns in completed returns. The real shift is not in the calculations that computers have handled for years, but in the research and advisory side. AI assistants can now answer complex tax questions on the spot, freeing preparers to spend less time digging through tax code and more time actually advising clients.

Success Story
GRF CPAs & Advisors

When *GRF CPAs & Advisors* in Bethesda, Maryland started experimenting with AI audits in 2017, most accounting firms were still debating whether the technology was ready. Four years later, they had become a case study in doing AI adoption right.

"I honestly think that the biggest risk is not doing anything," says partner Ricardo Trujillo.

Their approach: Begin with ***MindBridge*** for transaction analysis, let partners choose which engagements to use it on, and document everything. The platform now analyzes every transaction rather than sampling—a fundamental shift in how audits are conducted.

Results:

- **15% of audit engagements now use AI.**
- **GRF was named to Accounting Today's Best Accounting Firms to Work For.**
- **Awarded Inside Public Accounting's Top 200 Firms.**

"It's taken us four years to get where we are today," says Trujillo, "and I think we have a greater advantage for what's coming down the pipeline."

Three Best AI Tools for Tax Preparation

1. *Taxdome* -- AI Auto-tags Documents and Answers Natural Language Queries

What it does: Practice management platform with AI-powered document tagging and organization, automated client workflows, secure client portal, e-signature capabilities, and automated communication.

Why tax preparers choose it: The platform eliminates manual document organization and client chasing, with AI automatically categorizing and renaming uploaded documents.

Real impact: Tax firms report 70% reduction in time spent organizing client documents. Automated workflows eliminate manual task tracking and follow-up.

Monthly cost: $50-75/team member

Best for: Tax preparation firms tired of manual document organization and client communication during tax season.

60-day milestone: Your client documents auto-organize by type and year, with automated reminders eliminating your document collection headaches.

2. *Drake Tax* -- Professional Tax Prep With Integrated AI Research Assistant

What it does: Professional tax preparation software with AI-enhanced error detection, automatic calculation checks, electronic filing, and prior-year data import.

Why preparers choose it: The software combines comprehensive tax preparation capabilities with AI that catches errors before filing, reducing rejections and amendments.

Real impact: Tax preparers using ***Drake Tax*** report significantly fewer amended returns due to AI error detection. Prior-year carryforward saves hours of manual data entry.

Monthly cost: $1,595/year for standard package (effectively $133/month during tax season)

Best for: Solo preparers and small firms seeking reliable tax software with AI error checking.

60-day milestone: Your error rate drops significantly with AI flagging calculation issues before you submit returns.

3. *Lacerte* -- Intuit's AI Builds Custom Tax-saving Plans From Returns

What it does: Advanced tax preparation software for complex returns with AI-enhanced diagnostics, multi-state capabilities, partnership and S-corporation features, and integration with practice management tools.

Why preparers choose it: The software handles complex tax scenarios with AI that identifies optimization opportunities and compliance issues automatically.

Real impact: CPA firms report comprehensive diagnostic features catch issues that save significant amendment work and client penalties.

Monthly cost: $4,000-8,000/year depending on features and return count

Best for: CPA firms handling complex business returns and high-net-worth individuals.

60-day milestone: Your complex returns include comprehensive diagnostics flagging optimization opportunities you might have missed manually.

Management Consulting and Coaching

Consultants and coaches are finding that AI handles the background work that used to pad their hours but never impressed clients. Transcribing meeting notes, drafting follow-up emails, researching, and pulling together presentation decks that used to take half a day. The bigger opportunity is in scaling personalized advice. AI can now analyze client data, spot patterns across your clients, and surface recommendations that would have taken a senior consultant hours to develop, letting consultants spend their time on the strategic conversations that actually move the needle.

Success Story
Strategic Consulting Group

For over eight years, *Elena** built her management consulting practice, serving mid-market companies. Her clients value her strategic insights, but research and analysis preparation consumed 60% of her billable time, which limited her capacity for high-value strategic work.

AI Implementation Strategy:

- **Notion AI** is utilized for research synthesis and strategic planning.
- ***Jasper AI*** is used for content and presentation development.
- ***Salesforce Einstein*** is a tool used for analyzing client relationships and identifying opportunities.

Results after 10 months:

- The time required for research and analysis decreased by 65%, with an increase in accuracy and richer data.
- **Client capacity was enhanced by 40%** due to improved operational efficiencies.
- Due to premium strategic positioning, **revenue grew 156%.**
- **Service prices increased 89%** through enhanced data interpretation capabilities.

Three Best AI Programs for Management Consulting

1. *Notion AI* - The Strategic Knowledge Management System

What it does: AI-powered research blends with your planning, keeps you organized, creates professional client materials, and helps you work better as a team.

Why consultants choose it: Notion AI transforms fragmented information into organized strategic insights while maintaining complete control over your professional content.

Real impact: Management consultants report a 65% reduction in research compilation time and a 50% improvement in the quality of deliverables.

Monthly cost: $20/month per user.

Best for: Business consultants and coaches needing enhanced research and knowledge management systems.

2. *Jasper AI for Strategic Content* - The Business Intelligence Amplifier

What it does: AI-powered strategic content development, market analysis, competitive intelligence compilation, and presentation creation.

Why strategic professionals choose it: Jasper AI combines content generation with business intelligence, designed specifically for consultants and advisors.

Real impact: Consulting firms report a 70% improvement in strategic content quality and an 80% reduction in initial report draft time.

Monthly cost: $125/month per user for advanced features.

Best for: Management consultants seeking content development and market analysis support.

3. ***Salesforce Einstein for Consulting*** - The Client Relationship Intelligence Engine

What it does: AI-powered client analytics, opportunity scoring, engagement prediction, project pipeline management, and strategic relationship insights.

Why relationship-focused consultants adopt it: This tool delivers advanced client insights that support relationship management skills that help you build, maintain, and improve connections with clients.

Real impact: Crediting better relationship insights, consulting practices report a 45% improvement in client retention and a 67% increase in follow-on project success.

Monthly cost: $150/month per user.

Best for: Growing consultancy practices focused on a growing client base and relationship development.

Architecture and Design Services

Architects and designers are **using AI to generate initial concept renderings in minutes instead of days**, iterate through dozens of layout variations based on client desires, and even ensure code compliance before a person ever reviews the plans. The technology is handling the repetitive drafting and revision cycles that used to burn through project budgets before the interesting work could begin.

Success Story
Austin Architectural

*Michael** runs a six-person architectural firm in Austin, Texas, which specializes in sustainable residential and small commercial projects. His firm was losing opportunities to larger practices that could deliver faster design iterations and more comprehensive sustainability analysis. He needed to implement a solution.

AI Implementation Strategy:

- ***Autodesk AI*** design assistance seamlessly integrates with current CAD workflows.
- ***Sefaira*** for AI-driven sustainability analysis.
- ***Monday.com*** with AI-powered project intelligence for construction administration.

Results after 8 months:

- **Design development time was cut by 45%** while boosting sustainability metrics.
- Client presentations now include real-time energy modeling and optimization.
- The project delivery **timeline has been shortened by 30%** thanks to AI-assisted architectural design coordination.
- **Revenue has grown by 78%** due to expanded capacity and premium sustainability services.

Three Best AI Programs for Architecture

1. *Autodesk Construction Cloud with AI Design Assistance* - The Architectural Intelligence Platform

What it does: AI-powered design optimization, building performance analysis, building code compliance, construction coordination, and collaborative design workflows.

Why architects choose it: Integrates AI capabilities with existing Autodesk workflows while maintaining design and creative vision control .

Real impact: Architectural firms report a 45% reduction in design development time while improving building performance and code compliance.

Monthly cost: $200/month per user.

Best for: Small to medium architectural firms focused on design efficiency and building performance optimization.

2. Sefaira with AI Sustainability Analysis - The Environmental Performance Intelligence Engine

What it does: AI-powered energy modeling, daylight analysis, thermal comfort optimization, carbon footprint calculation, and sustainability reporting.

Why sustainability-focused architects adopt it: Sefaira provides sophisticated environmental analysis that enhances design decisions and client presentations.

Real impact: Architectural firms report a 60% improvement in sustainability analysis accuracy and 40% faster energy modeling workflows.

Monthly cost: $89/month per user.

Best for: Architects who are focused on sustainable design and green building certification.

3. Procore with AI Project Intelligence - The Construction Administration Brain

What it does: AI-enhanced project management, construction administration, document management, progress tracking, and client and subcontractor communication.

Why architects choose it: Procore streamlines construction administration while maintaining design intent and quality control throughout the building project.

Real impact: Architectural practices report a 50% reduction in construction administration time while improving project delivery outcomes.

Monthly cost: $24/month per user (architecture-specific features).

Best for: Architects providing construction administration services who need comprehensive project coordination tools.

Connect: Making AI and Professional Expertise Work Together

The transformation of professional services occurs when the analytical capabilities of AI are integrated with human expertise, resulting in advisory solutions that surpass what either could accomplish independently.

Legal Research Enhancement

AI Analysis: ***CaseText*** analyzed thousands of immigration cases and identified precedents relevant to a family's asylum application, including recent circuit court decisions and policy changes.

Human Expertise: Attorney David Thompson interpreted AI findings within the context of the client's specific circumstances, cultural background, and case strategy. ***CaseText*** helped develop arguments that addressed immigration officer concerns, which were relevant to the case.

Enhanced Result: Comprehensive case preparation resulted in a 60% reduction in approval time while addressing legal complexities not identified by automated systems. This approach often led to successful outcomes and increased client referrals to the immigration law network.

Accounting & Bookkeeping: Advisory Transformation

AI Handles: Transaction categorization, bank reconciliation, invoice processing, receipt data extraction, and financial statement generation.

People Provide: Financial analysis, tax strategy, business advisory, exception handling, and complex accounting judgment.

Working Together Results: Accountants transition from compliance reporting to strategic advisory services. Firms increase revenue per client while reducing time spent on routine bookkeeping.

Tax Preparation: Season Efficiency Without Burnout

AI Handles: Document organization, data entry, error detection, calculation verification, and client communication automation.

People Provide: Tax strategy, optimization planning, complex return preparation, client education, proactive planning.

Working Together Results: Tax season becomes manageable without 80-hour weeks. Preparers focus on complex returns and strategic planning rather than data entry.

Architectural Design Enhancement

AI Analysis: ***Sefaira*** modeled energy performance, daylight optimization, and thermal comfort for residential projects, automatically testing thousands of design variations.

Human Vision: Architect Michael applied AI insights to design decisions that balanced sustainability goals with client aesthetic preferences and budget constraints.

Collaborative Result: A LEED Platinum design was achieved, delivering a 30% better energy performance than code requirements while meeting the client's vision and budget, resulting in three referrals and industry recognition.

Strategic Consulting Integration

AI Research: Processing data from hundreds of sources, ***Notion AI*** compiled market analysis and intelligence for a client's expansion strategy.

Human Strategy: Consultants interpreted AI insights within the context of their client's organizational culture, financial constraints, and market positioning to develop actionable strategic recommendations.

Superior Outcome: Developed a data-driven strategic plan that **achieved a 234% improvement in market entry success** while maintaining the client's operational values and capabilities.

Keep: Maintaining Professional Excellence

Professional service firms that effectively implement AI develop advantages that grow stronger over time. Building systems that improve advisory capabilities while maintaining the client relationships that are central to their value.

Operational Efficiency and Competitive Positioning Through AI

By leveraging ***Salesforce AI*** features, the firm achieved substantial operational efficiencies that translated into improved client outcomes. These efficiencies were realized without sacrificing the high standard of personal service that clients expect, even amidst rapid technology adoption. Throughout this process, the firm maintained its commitment to exceptional service quality, ensuring that technology complemented rather than replaced the human element in client interactions.

Additionally, the integration of **AI into the advisory processes contributed to establishing a strong competitive position in the market.** The enhanced advisory capabilities enabled by AI allowed the firm to deliver more insightful and value-added guidance, thereby differentiating its services from competitors and reinforcing its reputation as a forward-thinking, client-focused professional service provider.

Professional Development Through Technology

Frame the implementation of AI in your organization as a demonstration of commitment to exceptional client service, not just a cost-cutting measure.

- **Use technological capabilities to provide more sophisticated analysis and strategic insights.**
- **Demonstrate how AI improves professional judgment and relationship management.**
- **Build a reputation as a professional who combines cutting-edge capabilities with trusted guidance.**

The Professional Services Advantage

The most successful professional service providers do not treat AI as a shortcut around expertise or a replacement for the judgment clients pay premium fees to access. They use it as a practice amplifier that requires experienced professionals to fully realize its potential.

Key Success Elements:

- AI handles research compilation, document preparation, data analysis, and routine client communications.
- Professionals provide strategic interpretation, complex problem-solving, ethical judgment, and relationship management.
- Client relationships remain personal while service capabilities expand beyond what human-only approaches can deliver.

- Professional standards stay tight because humans review everything before it reaches clients or regulatory bodies.

Professional services excellence is not about choosing between deep expertise and modern efficiency. **It is about using AI to make experienced professionals more thorough, more responsive, and more valuable than ever before.**

Practices thoughtfully using AI create advantages that automated platforms and offshore alternatives cannot replicate. **The combination of comprehensive analysis, rapid delivery, and trusted personal guidance takes years to develop.** A legal research database is just information. An immigration attorney who can interpret that research in the context of a family's specific circumstances, who understands the nuances that determine whether a case succeeds or fails, who guides anxious clients through a process that will change their lives? That is expertise no algorithm can provide.

GRF CPAs did not become less rigorous when they implemented AI audit tools. **They became a firm that analyzes every transaction instead of sampling, catching anomalies that traditional methods would miss.** Four years of patient implementation put them ahead of competitors who are still debating whether to start. Elena did not lose her strategic edge when she added AI research tools. She cut her analysis time by 65% and increased her capacity by 40%, which meant more time for the high-value strategic conversations that clients actually hire her for. Michael's architecture firm did not sacrifice design vision when they implemented sustainability modeling. They started winning projects that larger competitors used to dominate, delivering LEED Platinum results that would have been impossible to analyze manually.

Here is what should keep professional service providers who wait to implement AI up at night: While you are deciding whether research automation is worth the learning curve, your competitor across town just delivered a comprehensive analysis in two days that would have taken you two weeks. You are still billing for research hours. They are billing for strategic insight.

The attorney who implemented AI legal research six months ago prepares briefs with precedents you would never have found manually. The consultant who automated her data synthesis presents market analysis that makes your PowerPoint look like guesswork. The architect who added energy modeling wins sustainability projects you did not even know how to bid. **They are not technology companies. They are professionals who decided that "thorough but slow" was no longer good enough when clients expect thorough and fast.**

Your clients have options. The automated legal services keep expanding. The offshore research firms keep improving. The AI-powered consulting platforms get more sophisticated every quarter. **The relationships you have built over years matter, but relationships do not survive indefinitely when deliverable quality falls behind what clients can get elsewhere.** The client who trusts your judgment will still take a meeting with the professional who delivered better analysis in half the time.

The professional licensing boards are watching. The competitive landscape is shifting. The client expectations are not reverting to what they were five years ago. **The only question is whether you will use AI to become the professional who combines decades of expertise with the analytical depth that modern clients demand, or whether you will keep telling yourself that reputation alone will protect you from competitors who offer reputation plus everything else.**

Your judgment is not being automated. Your professional ethics are not being outsourced. Your ability to guide clients through complex decisions that require human wisdom is not something a platform can replicate. But the professionals who wrap that expertise in AI-enhanced analysis will win the engagements. They will command the premium fees. They will build the practices that compound over time while everyone else wonders where the referrals went.

Ricardo Trujillo at *GRF* said it well, "I honestly think that the biggest risk is not doing anything." Four years of disciplined AI adoption put his firm on Accounting Today's Best Firms list while competitors were still waiting for the perfect moment. **That perfect moment never comes. The professionals who start now, learn fast, and build systematically will have advantages that late adopters cannot catch.**

Your expertise is not obsolete. But expertise alone is no longer enough. The professional services firms winning today are not replacing human judgment with automation. **They are using AI to make their judgment more powerful, their analysis more comprehensive, and their client relationships stronger than human-only or AI-only approaches can achieve.**

How long before your best clients start wondering why the other professional always seems to know more and deliver faster?

Conclusion

The AI-Enhanced Future

The conversation that started this book, the one happening in kitchens, coffee shops, and offices across America, ends here with clear determination.

You do not have to choose between embracing AI and staying true to your values. No need to sacrifice the personal relationships that built your business to stay competitive. And better yet, you will not need to become a technology company to benefit from artificial intelligence.

What you need to do is what other business owners in this book did and use the framework set out to implement AI. **Make your expertise more powerful, your customer relationships stronger, and your business more valuable.**

The team at *Marten Machining* did not stop being precision craftsmen when they added AI-enhanced CNC programming. They became craftsmen who could tackle exacting jobs that manual methods could never achieve.

Poppy's Pizza & Grill did not lose their family atmosphere when it implemented AI phone systems. They created more time to focus on food and their famous hospitality.

Henderson & Associates in Austin did not lose touch with clients when they let ***QuickBooks AI*** handle expense categorization and invoice reminders. They stopped drowning in data entry and started having actual advisory conversations.

Insurance agencies that implemented AI-powered client communication did not become impersonal. **Their team finally had time to call customers before renewal season** instead of just processing paperwork.

The pattern is clear across every industry: businesses that implement AI correctly do not replace what makes them special; they magnify what makes them unique.

The shift happening across professional services tells the story clearly. **AI is handling the grunt work that used to steal hours from time spent with clients**: sorting documents, chasing down missing paperwork, categorizing transactions, and preparing routine communication. That is freeing professionals to do what they actually got into business to do. Work with clients and build relationships that matter.

AI is changing fast. The specific tools mentioned in this book will evolve. Some will merge with competitors. Others will add features we cannot predict. A few may disappear entirely.

That is precisely why this book focuses on framework and principles rather than step-by-step software tutorials. The TRACK methodology works whether you are implementing ***QuickBooks*** AI today or whatever replaces it in three years. The relationship-first philosophy applies regardless of which scheduling tool dominates your industry in the next decade.

Business owners who wait for AI to "settle down" will wait forever. **The ones succeeding right now are not technology prophets. They are pragmatists who picked a reasonable starting point, learned from implementing it, and built the confidence to adapt when better options emerged.**

Your first AI tool does not need to be your last. It just needs to be your first.

Your competitive **window is still open,** but it is narrowing. The numbers do not lie: Businesses using AI-powered invoice reminders **get paid 45% faster**. Firms with automated document processing handle more clients without adding staff. Every month you delay, another competitor gains an advantage that becomes harder to match. **Each day of hesitation represents a missed opportunity** to implement AI solutions that both protect your business relationships and enhance capabilities beyond those of your competitors.

So what is stopping you? Not the technology. Not the cost. Not your lack of a computer science degree. If any of those were real barriers, the business owners in this book would never have succeeded. **The only barrier left is your decision to begin.**

The TRACK Framework provides you with a roadmap to ensure you choose the right AI applications. The industry examples show you how AI works in your specialty. The only question remaining is whether you will use them.

Your **customers are expecting more clarity, speed, and better customer service.** Your **employees want work that matters and systems that make them more productive. Your community is watching to see whether you lead with intention or hesitate and let change happen *to* you.**

Stop waiting for conditions to be perfect. **Chasing perfection can prevent you from acting and achieving something great. Successful business executives are those who decide, test, adjust, and move forward even when conditions are not ideal.**

AI is not a replacement for what you have built. **AI is a tool to strengthen, protect, and make your business more resilient.**

Your future is not five years away. **The future is available now** to leaders willing to act with clarity and discipline. You have done the hard work already. It is time to embrace the opportunities ahead.

APPENDIX A

Businesses Referenced in AI for Real Companies

The following table catalogs every business featured in this book, organized by order of appearance. Businesses marked with an asterisk (*) are composite characters or have had identifying details changed to protect privacy.

Businesses appear in non-bold italics throughout the manuscript text.

Business Name	Contact / Staff	Title / Role	Location	Ch.	Pg.
LL Flooring Solutions	N/A (Corporate)	President & CEO	Richmond, VA	Intro	5
Badger CPA	Shannon Badger	Owner/CPA	San Antonio, TX	3	54
*Paul's Catering**	Paul*	Owner	Not Specified	5	83
*Johnson's Auto Repair**	Mike Johnson*	Owner	Suburban Location	7	103
*Tom Simpson's Hardware Store**	Tom Simpson*	Owner	Not Specified	7	106
Kaplan Early Learning Company	N/A (Corporate)	N/A	Lewisville, NC	7	108
Foothills Roofing & Exteriors	Armando Jacox Jr. & Annie Jacox	Co-Founders	Denver, CO	7	112
Suzanne Sansom Insurance	Suzanne Sansom	Owner/Agent	Katy, TX	7	114
Trio Heating and Air	David Katz & Michael Katz	Co-Founders	San Jose, CA	7	115
Paumanok Veterinary Hospital	Dr. David Hensen & Dr. Deirdre Hensen	Co-Owners/Veterinarians	Patchogue, NY	7	117
Oath Pizza	Max Seel	Co-Founder	Nantucket, MA	8	124
*Keith's Insurance Agency**	Keith*	Owner	Washington State	8	125

Business Name	Contact / Staff	Title / Role	Location	Ch.	Pg.
*Rodriguez Marketing Agency**	Carlos Rodriguez*	Owner	Not Specified	8	126
Pet Supplies Plus	N/A	Grooming Dept.	N/A	8	127
*Baldwins' Accounting & Tax Prep**	Terry Baldwin*	Owner	Not Specified	9	131
*Glori's Landscaping**	Glori*	Owner	Not Specified	9	134
TBH Complete Auto Care	Benjamin Hart	Owner	San Diego, CA	10	141
*Roberto's Marketing Agency**	Roberto*	Owner	Not Specified	10	142
*Thompson's Landscaping**	Mike Thompson*	Owner	Not Specified	11	148
Marten's Machining	Al Marten	Founder/Owner	Stevens Point, WI	13	198
Desco Professional Builders	N/A	N/A	Ellington, CT	13	208
Heritage Farm	Family-Owned (3rd gen.)	Owners	Central Illinois	14	229
Bles Dairies Farm	N/A	N/A	Netherlands	14	234
Gamble Estates Winery	Tom Gamble	Owner/Vintner	Napa Valley, CA	14	239
Best Quality Electric (BQ Electric)	Brian Quinn	Founder/Owner	Hicksville, Long Island, NY	15	253
*Marsh Cleaning**	Cynthia*	Owner	Portland, OR	16	271
Top Shop Auto	Frank Powell	Owner	Lafayette, CA	16	275
Synergy Dance Academy	Kari Fisher	Owner	Not Specified	16	278

Business Name	Contact / Staff	Title / Role	Location	Ch.	Pg.
Windows & Wheels Auto Detailing	N/A	N/A	Not Specified	16	281
Mission Auto KC	N/A	Owner	Kansas City, MO	16	286
Salon 77 West	N/A	Owner	Pleasanton, CA	16	288
Fit4MOM	N/A (Franchise)	Franchise Owners	Multiple Locations	16	294
MacDougalls' Cape Cod Marine Service	N/A	Owner	Falmouth Harbor, MA	16	296
Edgartown Harbor	N/A	Harbor Master	Edgartown, MA	16	297
MakerFlo	John Modi	Founder	Not Specified	17	304
*Rodriguez Gifts**	Elena Rodriguez*	Owner	Not Specified	17	306
CarGari	Rafael Small	Founder	Colorado	17	310
SodaPup	N/A	N/A	Not Specified	17	313
ThriftBooks	N/A	N/A	Not Specified	17	315
*Community Psychiatry / Dr. Sarah Che**	Dr. Sarah Che*	Owner/Therap ist	Portland, OR	18	322
Dr. Matthew Hitchcock's Practice	Dr. Matthew Hitchcock	Practitioner	Not Specified	18	327
Poppy's Pizza & Grill	N/A	Owner(s)	Estes Park, CO	19	340
*Santos Design Studio**	Maria Santos*	Owner/Creativ e Director	Portland, OR	20	359
*Content Agency**	N/A	N/A	Denver, CO	20	368
Thompson Marketing Services	N/A	N/A	Boston, MA	20	369
*Digital Agency**	N/A	N/A	Not Specified	20	371

Business Name	Contact / Staff	Title / Role	Location	Ch.	Pg.
*Topper's Transportation**	Mike Topper*	Owner	Texas/Oklahoma	21	388
Glo Pals	N/A (Mississippi State origin)	N/A	Not Specified	21	394
The DeBerry Team	Roxanne DeBerry	Founder/Owner	Plano, TX	22	406
Carson Group	N/A (Corporate)	N/A	National	23	427
RFG Advisory	N/A	N/A	Not Specified	23	430
Riverbank Finance	Anthony Bird & Joshua Woodwyk	Co-Founders	Grand Rapids, MI	23	434
GRF CPAs & Advisors	N/A	N/A	Bethesda, MD	23	439
Sisu Legal	N/A	N/A	Not Specified	23	443
*Elena's Management Consulting**	Elena*	Owner/Consultant	Not Specified	24	455
*Austin Architectural**	Michael*	Owner/Architect	Austin, TX	24	462
*Henderson & Associates**	N/A	N/A	Austin, TX	Conclusion	478

** Indicates composite character or pseudonym used to protect business identity.*

APPENDIX B

AI Tools Referenced in AI for Real Companies

The following table catalogs every AI tool and software platform referenced in this book, organized alphabetically. Tools appear in bold italics throughout the manuscript and are featured in the "The Three Best…" recommendation lists within each industry chapter.

AI tools appear in bold italics throughout the manuscript text.

AI Tool	Category / Type	URL	Ch.	Pg.
10Web AI Builder	AI Website Builder	10web.io	20	372
AdCreative.ai	AI Ad Generation	adcreative.ai	20	376
Adobe Creative Suite / Sensei	Creative Design Suite	adobe.com	20	360
Adobe Premiere Pro / Sensei	Video Editing AI	adobe.com	20	380
AI-powered Product Studio	Product Photography AI	google.com	17	312
Alexa (Amazon)	Voice AI Assistant	amazon.com/alexa	1	14
Applied Epic AI	Insurance Agency Mgmt	appliedsystems.com	7, 23	118, 437
Autodesk Construction Cloud	Architecture AI	construction.autodesk.com	24	464
Autodesk Fusion 360 / AI Design	CAD/CAM Manufacturing	autodesk.com	13, 24	200, 464
Bill.com	AP/AR Automation	bill.com	23	441
Birdeye	Review Management	birdeye.com	7	113
Blend	Digital Lending Platform	blend.com	23	436
Blueprint AI	Therapy Treatment Plans	blueprint-health.com	18	330
Cabinet Vision	Cabinet Design/Mfg	cabinetvision.com	13	213
Campspot	Campground Management	campspot.com	16	298

AI Tool	Category / Type	URL	Ch.	Pg.
Canva / Canva AI	Graphic Design	canva.com	Intro, 20	7, 362
Canva Pro / AI Magic Tools	Design Platform	canva.com	20	362
CaseText AI	Legal Research AI	casetext.com	24	450
CattleEye	Livestock Lameness Detect.	cattleeye.com	14	236
Chatfuel	Chatbot Builder	chatfuel.com	1	14
ChatGPT	General AI Assistant	openai.com	Intro, 1	4, 15
Claude (Anthropic)	General AI Assistant	claude.ai	Intro	4
Clio with AI / Clio Legal AI	Legal Practice Mgmt	clio.com	12, 24	168, 449
Connecterra IDA	Dairy AI Assistant	connecterra.io	14	235
Copy.ai	AI Copywriting	copy.ai	20	369
Creatopy AI	Display Ad Creation	creatopy.com	20	377
DALL-E	AI Image Generation	openai.com/dall-e	20	358
Descript	Audio/Video Editing AI	descript.com	20	379
Digitail	Veterinary Practice Mgmt	digitail.io	7	117
Dockwa	Marina Management	dockwa.com	16	296
Drake Tax	Tax Preparation AI	drakesoftware.com	23	442
eMoney Advisor	Financial Planning	emoneyadvisor.com	23	430
EOSDA Crop Monitoring	Satellite Crop Intel	eos.com	14	232
EZLynx / EVA Virtual Assistant	Insurance Agency Mgmt	ezlynx.com	7, 23	114, 435
FabriWIN	Fabrication Scheduling	fabriwin.com	13	217
Figma AI	UI/UX Design AI	figma.com	20	373

AI Tool	Category / Type	URL	Ch.	Pg.
FinishMatch AI	Wood Finish Matching	finishmatch.com	13	212
Fishbowl Manufacturing	Inventory Management	fishbowlinventory.com	12	167
Floify	Mortgage POS Automation	floify.com	23	435
Follow Up Boss	Real Estate Lead Mgmt	followupboss.com	22	414
Freightos	AI Freight Matching	freightos.com	21	399
Fresha	Beauty Booking Platform	fresha.com	16	290
Gavel AI (fmr. Documate)	Legal Doc Automation	gavel.io	24	451
GlossGenius	Beauty Business Mgmt	glossgenius.com	16	291
Google Ads Performance Max	AI Ad Optimization	ads.google.com	17, 20	311, 377
Google Analytics 4	Web Analytics	analytics.google.com	17	311
Google Gemini	General AI Assistant	gemini.google.com	Intro	4
Google My Business Insights	Local SEO Analytics	business.google.com	7	113
Google Translate	Translation AI	translate.google.com	1	14
Groomer.io	Pet Grooming Software	groomer.io	8	127
Harvey AI	Legal AI Assistant	harvey.ai	1	19
HawkSoft	Insurance Agency Mgmt	hawksoft.com	23	438
Hootsuite / Hootsuite AI	Social Media Mgmt	hootsuite.com	12	166
HubDoc (Xero)	Document Collection AI	hubdoc.com	23	442
HubSpot Breeze / HubSpot CRM	CRM & Marketing	hubspot.com	7	108

AI Tool	Category / Type	URL	Ch.	Pg.
Jackrabbit Dance	Studio/Activity Mgmt	jackrabbitclass.com	16	292
Jasper AI	AI Content Writing	jasper.ai	20, 24	364, 458
Jobber	Home Service Business Mgmt	getjobber.com	9, 16	134, 272
JobNimbus	Construction CRM	jobnimbus.com	7	112
Lacerte	Tax Preparation Software	lacerte.intuit.com	23	443
Lofty (fmr. Chime CRM)	Real Estate CRM AI	lofty.com	22	412
MachineMetrics	Manufacturing IoT/AI	machinemetrics.com	12, 13	169, 202
Mastercam	CNC Programming	mastercam.com	13	203
Med-PaLM	Medical AI	google.com	1	19
Microsoft Copilot	General AI Assistant	copilot.microsoft.com	Intro	4
Microvellum	Cabinet/Woodwork Software	microvellum.com	13	209
Midjourney	AI Image Generation	midjourney.com	20	358
Mindbody	Fitness/Wellness Mgmt	mindbodyonline.com	16, 18	295, 329
MindBridge	AI Audit Analytics	mindbridge.ai	23	440
Mitchell 1 ProDemand / Manager SE	Automotive Diagnostics	mitchell1.com	10, 16	142, 284
Monarch AI Tractor	Autonomous Tractor	monarchtractor.com	14	240
Monday.com	Project Management AI	monday.com	24	463
Mortgage Coach	Visual Loan Presentations	mortgagecoach.com	23	437
Nitrogen (fmr. Riskalyze)	Risk Tolerance Platform	nitrogenwealth.com	23	431
Notion AI	Knowledge Management AI	notion.so	24	456
OneSoil	Precision Agriculture	onesoil.ai	14	231

AI Tool	Category / Type	URL	Ch.	Pg.
OpenTable with Guest Intel	Restaurant Reservations	opentable.com	19	347
Perplexity	AI Search Engine	perplexity.ai	Intro, 1	4, 15
Pike13 AI	Fitness/Studio Mgmt	pike13.com	16	294
Plainsight	Computer Vision QA	plainsight.ai	13	218
PlanGrid	Construction Documents	plangrid.com	15	260
Plantix	Pest & Disease Detection	plantix.net	14	230
Podium	Customer Messaging	podium.com	12	168
Popmenu AI Answering	Restaurant Phone AI	popmenu.com	19	341
Procore	Construction Project Mgmt	procore.com	15, 24	260, 466
QuickBooks Advanced	Advanced Accounting	quickbooks.intuit.com	12	170
QuickBooks AI / Intuit Assist	Accounting & Finance	quickbooks.intuit.com	Intro, 12, 23	6, 170, 440
Raken	Field Reporting	rakenapp.com	15	260
Receipt Bank (Dext)	Expense Management	dext.com	12	170
Restb.ai	Real Estate Image AI	restb.ai	1	19
Resy	Restaurant Reservations	resy.com	19	345
Roomvo AI	Visual Merchandising	roomvo.com	Intro	5
Route4Me	Route Optimization	route4me.com	21	396
RoverPass	RV Park Reservations	roverpass.com	16	299
Runway ML	Video Effects & Generation	runwayml.com	20	381

AI Tool	Category / Type	URL	Ch.	Pg.
Salesforce Essentials / Einstein	CRM & Sales AI	salesforce.com	7, 24	110, 459
Salesforce Manufacturing Cloud	Manufacturing CRM	salesforce.com	12	167
Samsara Fleet Management	Fleet Management AI	samsara.com	21	398
Sefaira	Sustainability Analysis AI	sefaira.com	24	465
ServiceTitan / ServiceTitan AI	Field Service Mgmt	servicetitan.com	7, 12, 15	115, 170, 255
Shopify Plus / Shopify AI	E-commerce Platform	shopify.com	17	308
Shopmonkey	Auto Shop Mgmt	shopmonkey.io	16	280
SigmaNEST	Sheet Metal Nesting	sigmanest.com	13	216
SimplePractice / AI Note Taker	Wellness Practice Mgmt	simplepractice.com	12, 18	168, 325
SmaXtec	Internal Health Monitoring	smaxtec.com	14	237
Square for Retail AI	Retail POS & Analytics	squareup.com	17	309
Structurely	Real Estate AI Chat	structurely.com	22	408
Surfer SEO	SEO Content Optimization	surferseo.com	Intro, 20	3, 371
Tails AI	Veterinary AI Assistant	tails.com	7	117
TaxDome	Tax Practice Mgmt	taxdome.com	12, 23	166, 441
Tekmetric	Auto Shop Analytics	tekmetric.com	16	284
Thermwood CNC	CNC Machining	thermwood.com	13	214
Toast POS / Toast.ai	Restaurant POS & Mgmt	toasttab.com	12, 19	169, 343
Top Producer	Real Estate CRM	topproducer.com	22	415

AI Tool	Category / Type	URL	Ch.	Pg.
Vagaro	Salon/Spa Booking & Mgmt	vagaro.com	16	288
VineSpring	DTC Wine Sales	vinespring.com	14	243
Vine View	Precision Viticulture	vineview.com	14	241
Webflow AI	AI Website Builder	webflow.com	20	372
Writesonic	AI Content Writing	writesonic.com	20	370
Zapier AI	Workflow Automation	zapier.com	Intro	6
Zocks	Meeting AI for Advisors	zocks.ai	23	428

URLs and pricing referenced in this book were accurate at the time of publication. Please verify current information directly with vendors.

INDEX

AI for Real Companies: A Practical Guide to Smarter Systems and Stronger Profits

Page numbers refer to the first significant mention of each term. Page ranges indicate extended discussion.

A

B

C

D

E

U

V

W

Z

www.ingramcontent.com/pod-product-compliance
Lightning Source LLC
LaVergne TN
LVHW100247210926
848480LV00003B/847
9798994797105